AF526996

Global Warming in 21st Century

Causes, Effects and Future

Global Warming in 21st Century

Causes, Effects and Future

by :

K. K. Singh

MD Publications Pvt Ltd
New Delhi
www.mdppl.com

Published by

MD Publications Pvt Ltd
"MD House", 11, Darya Ganj,
New Delhi-110002
Ph. : +91-11-41563325, 41562846
E-mail : **contact@mdppl.com**
Website : **www.mdppl.com**

© 2008, All rights are strictly reserved with the Publisher

ISBN : 978-81-7533-107-5

PRICE : Rs. 1095.00

Published and Printed by
MD Publications Pvt Ltd at Printext,
D2/149A, Jeewan Park, Uttam Nagar, New Delh.

PREFACE

Governments of the world have been grappling with the problem of global warming for over two decades. Warming of the Earth's atmosphere is being increased by human activities particularly the burning of coal, oil, and other fossil fuels resulting in the emissions of carbon dioxide and other harmful "greenhouse gases" (GHGs). Global warming in turn is causing climate change, which is manifested in rising sea levels, droughts and floods, damage to agriculture, and harm to natural ecosystems and species.

Policy makers increasingly find that debates over environmental standards have become globalized, to borrow a word that has come into fashion in several contexts. Global warming is the most prominent of those issues: Indians now confront claims that the types of cars they choose to drive, the amount and mix of energy they consume in their homes and factories, and the organisation of their basic industries all have a direct effect on the lives of citizens of other countries and, in some formulations, may affect the future of the planet itself.

Climate scientists have reduced the magnitude of predicted warming, suggesting milder future climate scenarios. Ecologists have shifted from predicting ecosystem collapse to predicting that net primary productivity will likely increase over the long run. And economists are no longer predicting large damages, but rather a mixture of damages and benefits.

These changes are so dramatic that it is not clear whether the net economic effects from climate change over the next century will be harmful or helpful. The new research further suggests that effects are likely to vary across the planet. We now expect temperate and polar countries to enjoy small economic gains, whereas tropical countries are more likely to suffer economic losses. The consequences of warming will vary across countries, the countries' interest in imposing controls will vary as well. Many countries will benefit from warming the very countries, ironically, that have contributed the most to historic emissions. The industrialized nations of the earth happen to lie in boreal and temperate climates, where warming is likely to prove beneficial. Countries in subtropical and, especially, tropical climates which to date have made no commitment to reduce their greenhouse-gas emissions are likely to be damaged by warming. As each country becomes aware of national impacts, the impacts will become more important to the countries and affect future negotiations about abatement measures and costs. Each country will perceive different rewards for itself in taking action, and that will make it increasingly difficult to construct international agreements. Successful agreements will almost certainly have to include a compensation package to encourage at least some nations to cooperate.

Many of the roots of the problem of global warming and climate change, as well as the available solutions to them, are firmly planted in Asia. Much as the world needs the United States and other developed countries to reduce their extensive emissions of GHGs, if this problem is to be addressed with adequate vigor. Other issues range from the management of forests, fisheries, and water resources to the preservation of species and the search for new energy sources. Not far in the background of all those new debates, however, are the oldest subjects of international politics competition for resources and competing interests and ideas concerning economic growth, the distribution of wealth, and the terms of trade. Thus, if we care about the problem of climate change, and if we care about the human suffering caused by it, we must seriously consider and understand the roles played by the countries of Asia.

Thanks to Mr. Pranav Gupta of **MD Publications Pvt Ltd,** New Delhi, who supported me fully for completing this book.

K. K. Singh

Contents

1

Global Warming: An Introduction

Introduction

Human activities are disturbing this long-standing balance in a fundamental manner, adding to concentrations of several greenhouse gases, including, according to Maskell and Mintzer, "Water vapor, the principal greenhouse gas which is expected to increase in response to higher temperatures that would further enhance the greenhouse effect".

Scientists who study the future potential of human-induced warming also point to several other natural mechanisms which could cause the pace of change to accelerate, through "biotic feedbacks," such as release of carbon dioxide and methane from permafrost and continental shelves in the oceans. The possibility of a "runaway" greenhouse effect by the year 2050 is raised in the literature, often with a palpable sense of urgency. This sense of urgency intensified at the turn of the millennium, as scientists traced the relationship between warming of the lower atmosphere and stratospheric cooling, a major factor in depletion of the ozone layer over the poles.

Evidence of Rising Temperatures

Scientific debates regarding global warming were taking place against a climatic background of rapidly rising surface global temperatures. According to records gleaned from ice cores, tree rings, and other "proxy indicators" of temperature, the 1990s was the warmest decade of the second millennium on the Christian calendar, with 1998 the warmest single year on record. Each of the 12 months from August 1997 through September 1998 set a new all-time worldwide monthly high-temperature record. Seven of the 10 warmest years in the past 130 years occurred during the 1990s. By 1998, the Earth had sustained 20 consecutive years above the 1961 to 1990 average temperature, with the upward curve steepening toward the end of the period.

Also, during 1998,

At least 56 countries suffered severe floods, while 45 baked in droughts that saw normally unburnable tropical forests go up in smoke from Mexico to Malaysia and from the Amazon to Florida. Spring in the Northern Hemisphere is coming a week earlier and the altitude at which the atmosphere chills to freezing is rising by nearly 15 feet a year.

The warmest temperatures (compared to averages) during 1998 occurred in North America, in a pattern that commonly occurs during El Nino years. Almost the entire world was much warmer than millennial averages during the year, as well. According to the National Aeronautics and Space Administration (NASA),

The El Nino, by itself, cannot account for either the observed long-term global warming trend or the extreme warmth of 1998. Because the Pacific Ocean temperature has returned to a more normal level, it is anticipated that the global temperature in 1999 will be less warm than during 1998, but will remain well above the long-term mean for the period of climatology, 1951–1980. The rapid global warming since the mid 1970s exceeds that of any previous period of equal length in the time of instrumental data.

The National Climatic Data Center said that the spring of 2000 was the warmest in 106 years of record keeping in the United States, averaging 55.5 degrees F., 0.4 degrees F. above the previous record, set in 1910. (January through May also was the warmest on record. Temperatures in the United States averaged 48.5 degrees F.; the old record, set in 1986, was 47.4 degrees F.) In the United States, all 50 states reported above-average temperatures, which is very unusual.

Worldwide, 2000 was the twenty-fourth consecutive March-through-May period in which temperatures were higher than average in the Northern Hemisphere. The effects of La Nina seem restricted to the tropics, where temperatures were slightly below average. Large areas of the tropical Pacific Ocean also were below average for the same reason. "It's quite amazing," said Kelly Redmond, Deputy Director and Regional Climatologist at the Western Regional Climate Center in Reno. Kelly said that La Nina, which continued into the year 2000 in the equatorial Pacific, produces shifts in sea surface temperatures and wind patterns that usually translate into cooler-than-normal temperatures for large areas of the country, especially northern latitudes. That pattern was broken during the first half of the year 2000. "For a La Nina year, it's been rather out of character," Redmond said.

An analysis of the climate of the last 1,000 years suggests that human activity is the dominant force behind the sharp global warming trend seen during the late-twentieth century. The analysis, by Thomas J. Crowley, a geologist at Texas A&M University, found that natural factors, such as fluctuations in sunshine or volcanic activity, were the most powerful influences on temperatures until 1900. After 1900, however, natural forces have accounted for only about a quarter of observed worldwide warming, Crowley estimates. "These twin lines of evidence provide further support for the idea that the greenhouse effect is already here," Dr. Crowley wrote in the July 14, 2000 issue Science.

Crowley fed past variations in the solar radiance, bursts of volcanic activity, and other natural variations into a computer model simulating the flow of energy to and from the earth. This model produced temperatures which matched most actual climate variations from the year 1000 to the middle 1800s. Crowley found that the same simulation broke down completely during the twentieth century. The only climatic "forcing" which remotely matched the jump in temperatures seen in the last half of the century was the rise in emissions of greenhouse gases.

Michael E. Schlesinger, a climatologist at the University of Illinois, said that policy makers and the public should not assume that temperature trends will follow a smooth course. He said the relationship between the oceans and the atmosphere is so complex that alterations in it could easily cause temporary cool periods or other unpredictable and sometimes abrupt changes that could create confusion and paralyze work to attack global warming.

A steady warming of the Earth during the twentieth century has been confirmed by readings taken in more than 600 holes drilled into Earth's surface, mainly at mining sites. Shaopeng Huang, Henry N. Pollack, and Po-Yu Shen took temperature readings from 616 previously drilled boreholes on six continents. The authors of this study write, "subsurface temperatures comprise an independent archive of past surface temperature changes that is complementary to both the instrumental record and imate proxies," such as tree rings. Underground measurements are more apt to detect long-term trends than many surface proxies, the authors of the borehole study assert.

The borehole data confirm that the warming trend accelerated in the latter half of the twentieth century. "Some 80 percent of that warming corresponds with the growth of industrialization," said Henry Pollack, a geology professor at the University of Michigan in Ann Arbor and co-author of the study in the February 17, 2000 issue of Nature. The study confirms others, and this indicates that the warming trend in the latter half of the twentieth century is without precedent in the past 400 to 1,000 years.

"We do not know of any combination of natural mechanisms that can explain this phenomenon," writes Jonathan Overpeck, a geoscientist at the University of Arizona, in Nature. "So we are left with the likelihood that human-induced global warming is underway, caused by emissions of 'greenhouse gases' such as carbon dioxide, and that the next century is going to see even greater warming," he added.

"The upper 500 meters of the crust is an archive of what has taken place over the past 1,000 years," said Pollack. "We send a thermometer down and take the temperature at a number of depths along the way, establishing a profile of temperature down the borehole". Borehole temperatures in Pollack's study also show greater warming over the past 500 years than temperatures calculated by other means. "My hunch is boreholes do a better job of establishing long-term trends," he said.

To a consensus of scientists who make up the Intergovernmental Panel on Climate Change (IPCC), a United Nations committee, global warming is no longer a matter of whether, but how much, how soon, and with how much damage to the Earth's flora and fauna, including humankind. Many of the scientists who are closest to the evidence are urging policy makers to decide what is to be done, and, failing remedies on a global scale, what price will be paid by future generations.

In 1997, a group of prominent scientists issued this warning: Our familiarity with the scale, severity, and costs to human welfare of the disruptions that the climatic changes threaten leads us to introduce this note of urgency and to call for early domestic action to reduce U.S. emissions via the most cost-effective means. We encourage other nations to join in similar actions with the purpose of producing a substantial and progressive global reduction in greenhouse-gas emissions, beginning immediately. We call attention to the fact that there are financial as well as environmental advantages to reducing emissions. More than 2000 economists recently observed that there are many potential policies to reduce greenhouse-gas emissions for which total benefits outweigh the total costs.

This statement was signed by George M. Woodwell, founder and director of Woods Hole Research Center; Dr. John P. Holdren; Teresa Heinz; John Heinz, Professor of Environmental Policy and Director, Program in Science, Technology, and Public Policy, John F. Kennedy School of Government; Dr. William H. Schlesinger; James B. Duke, Professor of Botany at Duke University; Jane Lubchenco, Ecology Professor at Oregon State University and chair of the American Association for the Advancement of Science; Dr. Harold Mooney, Paul S. Achilles Professor of Environmental Biology, Stanford University; Dr. Peter Raven, Director of the Missouri Botanical Garden; and F. Sherwood Rowland, professor of chemistry at the University of California at Irvine and recipient of the 1995 Nobel Prize in Chemistry.

Global warming has been no secret to atmospheric scientists. Articles on the subject began to appear occasionally in the scientific literature of meteorology and other fields during the middle 1970s. A scientific consensus was forming as early as October, 1985, at a conference in Villach, Austria, where a consensus statement began

As a result of the increasing concentrations of greenhouse gases, it is now believed that in the first half of the next century a rise in global mean temperature could occur which is greater than any in man's history.

Other conferences on global warming were convened in Villach and Bellagio, Italy. An executive summary of these meetings said that "It is now generally agreed that if present trends of greenhouse-gas emissions continue during the next hundred years, a rise in global mean temperature could occur that is larger than any experienced in human history". This summary also projected that the most extreme increases would come in the northern latitudes during the winter, a forecast that was being borne out by observations a decade later.

History of the Greenhouse Effect

The fossil-fueled industrial revolution was born in England. As coal-fired industry (as well as home heating and cooking) filled English skies with acrid smoke, some English homeowners protested coal's use as a fuel. Others described the horrors of coal mines, into which children as young as six years of age were sent to work. Queen Elizabeth sometimes forbade the burning of coal in London while Parliament was in session. In 1661, John Evelyn wrote a book that complained about the noxious nature of coal smoke.

The coal-burning steam engine was invented by Thomas Newcomen in 1712 and refined into a form which was widely adaptable for industrial processes by James Watt, beginning in 1769. Within a century of industrialism's first stirrings, during the 1820s, Jean Baptiste Joseph Fourier, a Frenchman, compared the atmosphere to a greenhouse. During the 1860s, John Tyndall, an Irishman, developed the idea of an "atmospheric envelope," suggesting that water vapor and carbon dioxide in the atmosphere are responsible for retaining heat radiated from the sun. Tyndall also wrote that climate might warm or cool based on the amount of carbon dioxide and other gases in the atmosphere. Tyndall, who speculated in 1861 that a fall in carbon dioxide levels could have accounted for the ice ages, was the first person to make quantitative, spectroscopic measurements showing that water vapor and carbon dioxide absorb thermal radiation and could therefore trap solar heat in the atmosphere.

In 1896, Svante August Arrhenius, a Swedish chemist, published a paper in The London, Edinburgh, and Dublin

Philosophical Magazine and Journal of Science titled: "On the Influence of Carbonic Acid in the Air upon the Temperature of the Ground." In his paper, Arrhenius theorized that a rise in the atmospheric level of carbon dioxide could raise the temperature of the air. He was not the only person thinking along these lines at the time; Swedish geologist Arvid Hogbom had delivered a lecture on the same idea three years earlier, which Arrhenius incorporated into his article. Arrhenius was a well-known scientist in his own time, not for his theories describing the greenhouse effect, but for his work in electrical conductivity, for which he was awarded a Nobel Prize in 1903. Later in his life, Arrhenius directed the Nobel Institute in Stockholm. His work in global-warming theory was not much discussed during his own life. Arrhenius, using the available measurements of absorption and transmission by water vapor and carbon dioxide, developed the first quantitative mathematical model of the Earth's greenhouse effect and obtained results of acceptable accuracy by today's standards for equilibrium climate sensitivity to carbon dioxide changes.

Arrhenius developed his theory through the use of equations, by which he calculated that a doubling of carbon dioxide in the atmosphere would raise air temperatures about 10 degrees F. Arrhenius thought 3,000 years would have to pass before human-generated carbon dioxide levels would double, a miscalculation which shares something with Benjamin Franklin's belief, late in the eighteenth century, that European-American expansion across North America would take a thousand years.

Arrhenius applauded the possibility of global warming, telling audiences that a warmer world "would allow all our descendants, even if they only be those of a distant future, to live under a warmer sky and in a less harsh environment than we were granted". In 1908, in his book Worlds in the Making, Arrhenius wrote, "By the influence of the increasing percentage of carbonic acid in the atmosphere, we may hope to enjoy ages with more equable and better climates, especially as regards the colder regions of the Earth, ages when the Earth will bring forth much more abundant crops than at present for the benefit of rapidly propagating mankind".

Arrhenius' ideas were not widely discussed, but they did not completely die during the early twentieth century. Alfred J. Lotka, an

American physicist, warned in 1924, Economically we are living on our capital; biologically, we are changing radically the complexion of our share in the carbon cycle by throwing into the atmosphere, from coal fires and metallurgical furnaces, ten times as much carbon dioxide as in the process of breathing.

Calculating on the basis of fossil fuel use in 1920, at the beginning of the automotive age, Lotka ventured that the level of carbon dioxide in the atmosphere would double in 500 years because of human activities, one-sixth of the time period forecast by Arrhenius.

By the late 1930s, the prospect of global warming was catching the eye of G.S. Callendar, a British meteorologist, who gathered records from more than 200 weather stations around the world. He argued that the Earth had warmed 0.4 degrees C. between the 1880s and the 1930s because of carbon dioxide emissions by industry. While Callendar 's assertions were met with skepticism by many English scientists at the time, he was laying the foundation for modern day efforts to make more precise measurements of atmospheric trace-gas trends and radiative properties and to design more capable climate models to simulate climate change.

Two decades after Callendar, in 1956, Gilbert Plass, a scientist at Johns Hopkins University in Baltimore, suggested that carbon dioxide is an important climate-control mechanism. He also projected that burning of fossil fuels would raise the global temperature 1.1 degrees C. (2.0 degrees F.) by the end of the century, very close to the actual worldwide increase.

"Grandfather of the Greenhouse Effect"

In 1957, Roger Revelle and Hans Suess warned, as part of the International Geophysical Year,

Human beings are now carrying out a large-scale geophysical experiment of a kind that could not have happened in the past nor be reproduced in the future. Within a few centuries we are returning to the atmosphere and oceans the concentrated organic carbon stored in sedimentary rocks over hundreds of millions of years.

Revelle would become known to the world some years later as the mentor of a graduate student, Albert Gore, who, in 1992 (a year after Revelle died), was elected vice president of the United States. The same year, Gore published a book, Earth in the Balance, which argued for mitigation of the greenhouse effect. Until the late 1950s, scientists had no reliable records of carbon dioxide and other greenhouse-gas levels in the atmosphere. At about the same time that Revelle and Suess issued their warning, Charles David Keeling began to assemble documentation indicating that the worldwide level of carbon dioxide had risen to about 315 parts per million compared to about 280 p.p.m. at the end of the previous century. Keeling 's calculations also indicated that the level was continuing to rise.

When Keeling decided to measure the concentration of carbon dioxide in the atmosphere, he first had to construct a machine to obtain readings in parts per million. No such machine existed at the time. Keeling worked on his "manometer" for a year, building the machine from an old blueprint at the California Institute of Technology in Pasadena. Keeling 's first readings, on the roof of a Caltech laboratory, showed an atmospheric concentration of 310 parts per million. Keeling next took his manometer on family vacations, recording readings. What he found contradicted scientific assumptions of the time, which held that carbon dioxide levels would vary widely, depending on local sources of the gas. Instead, he found that carbon dioxide readings were very similar in different places.

"I decided all the data in the literature was wrong," Keeling later told William K. Stevens, a science reporter for the New York Times. After Keeling and his associates had taken a number of readings for more than a year, Keeling made a discovery which confirmed Revelle and Suess' assertions. The readings on his manometer were rising steadily, year by year. The resulting graph which plots the level of carbon dioxide in the atmosphere has come to be known among scientists as "The Keeling Curve."

Keeling also was discovering that atmospheric carbon dioxide levels vary annually, with lower readings in the spring and summer (when plants are respiring oxygen on the large land masses of the Northern Hemisphere) and higher levels during fall and winter, when many plants are dormant and more carbon is being released due to vegetative decay. This annual cycle can vary by as much as three percent in the Northern Hemisphere, compared to one percent in the Southern Hemisphere, where the dominance of oceans mutes seasonal cycles and restricts variability in the carbon dioxide level.

Until the work of Revelle, Suess, and Keeling, most scientists who studied carbon dioxide levels in the atmosphere assumed that the oceans absorbed all of the extra carbon that human activities were injecting into the air. The readings of Keeling and his associates showed that human activity was steadily raising the proportion of carbon dioxide in the atmosphere more quickly than the oceans or other "sinks" of the gas could absorb it.

By the early 1960s, a growing number of scientists were watching as Keeling's carbon dioxide readings from the Mauna Loa observatory climbed steadily. In 1963, the Conservation Foundation issued a warning in a report titled Implications of Rising Carbon Dioxide Content of the Atmosphere. The report concluded: "It is estimated that a doubling of carbon-dioxide content in the atmosphere would produce a temperature rise of 3.8 degrees".

Revelle, who became known as the "grandfather" of the greenhouse effect, was born in Seattle on March 7, 1909, into a family of Huguenot descent on his father's side and Irish descent on his mother's side. His parents, William Roger Revelle, an attorney and schoolteacher, and Ella Robena Dougan Revelle, also a schoolteacher, both graduated from the University of Washington.

After he was admitted to Pomona College at the age of 16, Revelle entertained thoughts of a career in journalism. However, under the influence of a charismatic professor, Alfred Woodford, Revelle became interested in geology and, after receiving his B.A. in 1929, spent a year of additional study with Woodford. Revelle entered graduate studies at the University of California, Berkeley, in 1930, under the tutelage of geologist George Louderback, who stimulated his interest in marine sedimentation. In 1936, Revelle completed his doctorate at the University of California.

Called to active duty in the U.S. Navy six months before the attack on Pearl Harbor, Revelle was assigned to oceanographic research applied to wartime needs as an officer in the U.S. Navy. Near the end of World War II, Revelle was assigned to Joint Task Force One, the military command supervising the first postwar atomic test on Bikini Atoll (Operation Crossroads). He organized the Crossroads scientific program, which included a study of the diffusion of radio nuclides in the atoll, as well as radioactivity's impact on marine life. What he learned through these studies, and others, made Revelle a lifelong opponent of nuclear weapons. After the war, in 1951, Revelle became director of the Scripps Institute of Oceanography.

In 1964, Revelle accepted an appointment as Richard Saltonstall Professor of Population Policy at Harvard University, where he served as Director of the Center for Population Studies until 1974. Revelle held the endowed chair at Harvard until 1978. Revelle also was sympathetic to the 1962 book Silent Spring by Rachel Carson, which argued that ecosystems were being disrupted by the use of halogenated hydrocarbon pesticides such as DDT.

Revelle played a key role in the creation, during 1970, of the Scientific Committee on Problems of the Environment of the International Council of Scientific Unions (ICSU). He also suggested the objective for the ICSU 's International Geosphere-Biosphere Program in 1986: "To describe and understand the interaction of the great global physical, chemical, and biological systems regulating planet Earth's favorable environment for life, and the influence of human activity on that environment".

Revelle served as president of the American Association for the Advancement of Science during 1974. In 1976, he returned to the University of California, at San Diego, to become a professor of science and public policy. He received the National Medal of Science in 1991 "for his pioneering work in the areas of carbon dioxide and climate modifications, oceanographic exploration presaging plate tectonics, and the biological effects of radiation in the marine environment, and studies for population growth and global food supplies." To a reporter asking why he received the medal, Revelle said, "I got it for being the grandfather of the greenhouse effect".

Global Warming: Political Issue

The modern debate over whether the earth is warming due to human activity began in policy circles during 1979, after a small number of well-known scientists reported to the Council on Environmental Quality, "Man is setting in motion a series of events that seem certain to cause a significant warming of world climates unless mitigating steps are taken immediately". At the same time, the National Academy of Sciences initiated a study of the greenhouse effect. Also during 1979, in the United States, the President's Council on Environmental Quality mentioned global warming: "The possibility of global climate change induced by an increase of carbon dioxide in the atmosphere is the subject of intense discussion and controversy among scientists".

An alarm regarding global warming was sounded in the British journal Nature, May 3, 1979: "The release of carbon dioxide to the atmosphere by the burning of fossil fuels is, conceivably, the most important environmental issue in the world today". At about the same time, a study conducted by a scientific team chaired by meteorologist Jule Charney estimated that doubling the carbon dioxide level in the atmosphere would raise the average global temperatures by about three degrees C., plus or minus 1.5 degrees C. Four years later, the United States Environmental Protection Agency released a report, "Can We Delay a Greenhouse Warming." A National Academy of Sciences report, also issued in 1983, stated, "We do not believe that the evidence at hand about CO2-induced climate change would support steps to change current fuel-use patterns away from fossil fuels".

Until the early 1980s, those who argued that infrared forcing had (or would) raise the temperature of the atmosphere near the Earth's surface had a statistical problem. Starting in 1940, until about 1975, average global temperatures actually fell slightly. After 1975, temperatures began a steady, accelerating rise, which made global warming a notable political issue by the late 1980s.

During 1985, Veerabhaadran Ramanathan, Ralph Cicerone, and their colleagues at the National Center for Atmospheric Research argued that trace gases other than carbon dioxide could be as dangerous vis-à-vis greenhouse warming as carbon dioxide. Also, by 1985, Claude Lorius was beginning to demonstrate that lower carbon dioxide levels in the atmosphere strongly correlated with low temperatures during the last ice age.

The potential impact of warming due to infrared forcing was raised again at United Nations sponsored conferences in Villach, Austria, during the middle 1980s. Shortly after these conferences, Senators David Durenberger of Minnesota and Albert Gore of Tennessee called for an international "Year of the Greenhouse" to raise the issue in public consciousness. Gore already had played a role in congressional hearings on the issue in 1982 and 1984, when he was serving in the House of Representatives.

As a political issue in the United States of America, global warming came of age during the notably hot summer of 1988. The year 1988 provided something of a wake-up call in the debate over global warming because it was the warmest year since reliable records had been kept in the middle of the nineteenth century. During 1988, 400 electrical transformers in Los Angeles blew out on a single day as temperatures rose to 110 degrees F. Two thousand daily temperature records were set that year in the United States. Widespread heat and drought caused some crop yields in the U.S. Midwest to fall between 30 and 40 percent. In Moscow, Russians escaping their hottest summer in a century flocked to rivers and lakes, and drowned in record numbers.

During 1988, Colorado Senator Timothy E. Wirth, whose hearings on global warming the previous winter had drawn little attention, played the weather card. He called another hearing, this time during the summer. As it happened, the hearing convened on a particularly hot, humid day in Washington, D.C. during which the high temperature reached a record 101 degrees F. At Wirth's hearing, James E. Hansen, head of the federal government's Goddard Institute of Space Studies, testified that the unusually warm temperatures of the 1980s were an early portent of global warming caused by the burning of fossil fuels and not solely a result of natural variation. Hansen's remarks became front-page news nationwide within hours.

Hansen also continued a running battle throughout the 1980s, during the Reagan and Bush administrations, to call the science of global warming as he saw it, despite repeated threats to the funding of the Goddard Institute. The Office of Management and Budget forced Hansen to censor the severity of his findings several times. The pressure was so intense that Hansen sometimes asked to testify as a private citizen rather than as a federal employee.

In the meantime, the scientific debate over global warming was intensifying. By the end of 1988, the United Nations General Assembly had approved the creation of the Intergovernmental Panel on Climate Change (IPCC). A year later, Hansen said that it was "time to cry wolf":

When is the proper time to cry wolf? Must we wait until the prey, in this case the world's environment, is mangled by the wolf's grip? The danger of crying too soon, which much of the scientific community fears, is that a few cool years may discredit the whole issue. But I believe that decision-makers and the man-in-the-street can be educated about natural climate variability. A greater danger is to wait too long. The climate system has great inertia, so as yet we have realized only a part of the climate change which will be caused by gases we have already added to the atmosphere. Add to this the inertia of the world's energy, economic, and political systems, which will affect any plans to reduce greenhouse gas emissions. Although I am optimistic that we can still avoid the worst-case climate scenarios, the time to cry wolf is here.

Hansen elaborated,

I said three things [in 1988]. The first was that I believed the earth was getting warmer and I could say that with 99 percent confidence. The second was that with a high degree of confidence we could associate the warming and the greenhouse effect. The third was that, in our climate model, by the late 1980s and early 1990s, there's already a noticeable increase in the frequency of drought.

Between June 27 and 30, 1988, as the Earth's warmest summer on record (to that time) was getting under way, more than 300 leaders in science, politics, law, and environmental studies gathered in Toronto at the invitation of Canada's government to address problems related to climate change, including prospects of global warming. A scientific consensus was forming around the idea that human activity already was altering the Earth's atmosphere at an unprecedented rate. A consensus statement issued by the Montreal climate conference asserted that "There can be a time lag of the order of decades between the emission of gases into the atmosphere and their full manifestation in atmospheric and biological consequences. Past emissions have already committed planet Earth to a significant warming".

The climate system is never actually in thermodynamic equilibrium. Rather, it is forever playing catch-up with the daily and seasonal variations of incident sunlight, as the ground tries to come into thermal equilibrium with changes in solar radiation. This is where the heat capacity of the ground, the atmosphere, and the ocean come into play, as well as the thermal opacity of the atmosphere, which regulates how readily heat energy from the ground can be radiated to space. The time it takes the system to get to a new equilibrium is expressed in terms of a time constant, or "e-folding" time, in other words, the length of time that it takes the system to reach approximately 63 percent of its final equilibrium temperature. Mathematically, it takes forever to reach "true" equilibrium; but practically, after a few e-folding times, the system can be said to be in effective equilibrium. The e-folding time of the atmosphere is a few months; the ocean mixed layer, a few years; and the total (deep) ocean, a few hundred years. The present climate has accumulated about 0.5 watts per meter squared of unrealized warming.

In 1989, Veerabhaadran Ramanathan stated,

The rate of decadal increase of the total radiative heating of the planet is now about five times greater than the mean rate from the early part of this century. Non-CO2 trace gases in the atmosphere are now adding to the greenhouse effect by an amount comparable to the effect of CO2 increase. The cumulative increase in the greenhouse forcing until 1985 has committed the planet to an equilibrium warming of about 1 to 2.5 degrees C.

Ramanathan continued,

The climate system cannot restore the equilibrium instantaneously, and hence the surface warming and other changes will lag behind the trace-gas increase. Current models indicate that this lag will range [from] several decades to a century. However, analyses of temperature records of the last 100 years as well as proxy records of paleoclimate changes indicate that climate changes can also occur abruptly instead of a gradual return to equilibrium as estimated by models. The timing of the warming is one of the most uncertain aspects of the theory.

At about the same time, Peter Ciborowski projected, "Within 50 years, we will be committed to a mean global temperature rise of 1.5 degrees C. to 5 degrees C. And if no attempt is made to slow the rate of increase, we could be committed to another 1.5 degrees C. to 5 degrees C. in another 40 years".

Debate regarding the greenhouse effect intensified when temperature readings in 1990 eclipsed the record warmth of 1988.

Later in the decade, 1995 became the warmest year, followed by 1997, and 1998. During most of these years, The Southern Oscillation weather pattern (called "ENSO" in climate-change literature) was an important factor in world weather. The ENSO involves a marked warming of the tropical ocean off the west coast of South America. The pattern occurred more often during the 1990s than at any time for the century and a half during which detailed worldwide weather records have been available.

Some atmospheric scientists have asserted that the ENSO was associated with a gradual warming of the lower atmosphere during most of the twentieth century. In 1996, Kevin E. Trenberth and Timothy J. Hoar of the National Center for Atmospheric Research pointed out that El Nino periods had occurred more frequently in the 1980s and 1990s than during the previous century.

During the 1992 presidential campaign in the United States, candidate Bill Clinton criticized the Bush administration's refusal to join in worldwide diplomatic efforts to reduce emissions of greenhouse gases. Clinton's first budget proposed a carbon tax, a measure that was quickly dropped under pressure from Republicans in Congress. In Congress, the carbon tax died in committee. Meanwhile, a Climate Convention signed by 161 countries at the Conference on Environment and Development in Rio de Janeiro, during 1992, contained a directive, in Article 2, favoring stabilization of greenhouse gases "at levels and on a time scale that do not produce unacceptable damage to ecosystems and that allow for sustainable economic development".

At about the same time several studies indicated that daily minimum temperatures had increased more rapidly during much of the twentieth century than daily maxima. Easterling reports that between 1950 and the mid-1990s, daily minima have increased at a rate of 1.86 degrees C. per century, while maxima have increased 0.88 degrees C. (less than half as much) during the same period. A study by Henry F. Diaz and Raymond S. Bradley supported climate-model forecasts that daily minimum temperatures will rise more rapidly than maxima in a warmer world. Diaz and Bradley, who studied temperature changes during the twentieth century at high elevation sites, wrote, "The signal appears to be more closely related to increases in daily minimum temperatures than changes in the daily maximum. The changes in surface temperature vary spatially, with Europe (particularly Western Europe) and parts of Asia displaying the strongest high-altitude warming during the period of record".

A number of legislative bodies in different parts of the world took initiatives to limit greenhouse-gas emissions soon after the memorably hot summer of 1988. During 1989, the Netherlands passed a National Environmental Policy Plan, which required a freeze on carbon dioxide emissions at 1989–1990 levels. The parliament of Norway decided to limit carbon dioxide emissions in that country to 1989 levels by the year 2000, with a decline in emissions mandated after that. The Vermont State Legislature enacted a law outlawing automobile air conditioners after 1993 unless a substitute was developed to replace chlorofluorocarbons (CFCs). On June 13, 1990, just before Germany was reunified, the West German Cabinet committed the country to a 25 percent reduction in greenhouse gases, based on 1987 levels, by the year 2005. Later, the unified government stood behind these limits, but added allowances for energy-inefficient industries in what had been East Germany.

Stephen H. Schneider's Global Warming: Are We Entering the Greenhouse Century? was one of the first popular treatments of global warming in book form. During the next several years, Schneider became a leading scientific voice in public debates over the issue. By the late 1990s, Schneider was a professor in the Department of Biological Sciences and a senior fellow at the Institute for International Studies, at Stanford University. He was honored in 1992 with a MacArthur Fellowship for his ability to integrate and interpret the results of global climate research through public lectures, seminars, classroom teaching, environmental assessment committees, media appearances, Congressional testimony, and research collaboration with colleagues. Schneider served as a consultant to several federal agencies as well as White House staff in the Nixon, Carter, Reagan, Bush, and Clinton administrations.

Schneider's Global Warming: Are We Entering the Greenhouse Century? ends with a passage that sounded rather prescient in the year 2000:

"Are we now entering the Greenhouse Century?" I asked in the subtitle of this book. It should be clear by now that I believe we've been in it for a while already, but admit that it will take a decade or so more of record heat, forest fires, intense hurricanes, or droughts to convince the substantial number of skeptics that still abound. Unfortunately, while the antagonists debate, the greenhouse gases keep building up in the atmosphere. I wonder what we will say to our children when they eventually ask what we did – or didn'tdo – to create the Greenhouse Century they will inherit.

Schneider wrote, "I strongly suspect that by the year 2000 increasing numbers of people will point to the 1980s as the time the global warming signal emerged from the natural background of climatic noise". When he was asked whether human activities had assumed a dominant role in climate change, Schneider said, "I'm not 99 percent sure, but I am 90 percent sure. Why do we need 99 percent certainty when nothing else is that certain? If there were only a 5 percent chance the chef slipped some poison in your dessert, would you eat it?". Schneider argues that everyone will be 100 percent certain that scientists' beliefs about climate change are accurate only when they observe dramatic, adverse changes in climate. By then, he asserts, it will be too late to take remedial measures.

In 1991, author John J. Nance, in What Goes Up: The Global Assault on Our Atmosphere, quoted Dr. Susan Soloman, an atmospheric chemist:

I can't go home and dump my garbage in my neighbor's backyard. The police would arrest me in five minutes. But [until they were banned] I could take a tank of chlorofluorocarbons, put it in my backyard, turn it on and let it go into the atmosphere all day long, and no one [could] stop me. Somehow that's very wrong.

By the year 2000, global warming was becoming a national political issue in some countries. One was Norway, where, on March 9, Prime Minister Kjell Magne Bondevik announced the resignation of his government after losing a vote of confidence in parliament. Bondevik 's became the first government in the world to fall as a result of issues related to global warming. The issue was his objection to the building of several natural gas fired power stations. The minority three-party coalition government, with just 42 seats in the 165-member parliament, was bitterly at odds with the parliamentary majority over whether to build the gas fired power plants to burn some of Norway's large stocks of natural gas. The government argued that the plants would release far too much carbon dioxide linked to global warming. The government wanted to put the gas power plants on hold until more efficient, cleaner technology is developed to make natural gas fired power plants pollution free.

The opposition, an alliance of conservatives and labor, wanted to build the plants anyway, on the grounds that there is no alternative to meet the demand for electricity. Norway currently generates almost all its electricity from hydropower, which emits no greenhouse gases. Further hydropower development is not popular in Norway because of its effects on the landscape, according to the Environment News Service.

Britain's Activist Role in Climate-Change Diplomacy

Britain's government has diplomatically nudged the United States on several occasions on the global-warming issue. The British government's involvement in global warming as a political issue dates to the tenure of Tory Prime Minister Margaret Thatcher, who had earned an undergraduate degree in chemistry. In 1997, as 160 nations prepared to negotiate the Kyoto Protocol, British Deputy Prime Minister John Prescott traveled to Washington, D.C. to advocate a stronger response to the problem by the United States. According to a report by the British Broadcasting Corporation,

Mr. Prescott avoided direct criticism of the Clinton administration. He said simply that he had told his hosts that cutting greenhouse gases was presently the most important issue facing the world and that acceptable targets had to be agreed upon. But the British Minister of the Environment, Michael Meacher, who also recently visited Washington, was less cautious. He said it as now time for the Americans to show greater leadership.

Britain's environmental secretary, John Gummer, called for an emissions reduction target of 50 percent, beginning with an end to all subsidies for oil and coal use, as Britain's Financial Times pinned the ineffectiveness of the Kyoto Protocol on resistance from established interests in the United States:

Right now the final protocol promises to be a sad affair given the power of the oil and coal lobby in the United States, which has still failed to meet its obligations under the Rio Treaty. It is therefore a shame that the U.S. government lacks the courage to admit that while some jobs in some industries may be lost, new jobs will more than compensate.

During March, 1999, the European Union called for a 15 percent reduction in carbon dioxide, methane, and nitrous oxide emissions by 2010. At its December 15, 1999 meeting, the European Parliament passed a resolution supporting speedy approval of the Kyoto Climate Protocols. The resolution called for a carbon tax, as well as for emission controls. The lack of legislative action on global warming in the United States has been, by contrast, notable by its absence. According to United Nations figures, carbon dioxide emissions in Britain fell roughly eight percent between 1990 and 1998, while emissions rose 10.7 percent in the United States, 9.5 percent in Japan, and 12 percent in Australia.

Scientific Consensus

The public policy debate regarding global warming has often conveyed an impression that scientists are hopelessly divided over the issue of whether human activities are warming the lower atmosphere. In actuality, a high degree of agreement has existed since the IPCC 's First Assessment was published in 1990. The IPCC 's first major report forecast widely varying temperature rises by region with an assumed doubling of carbon dioxide in the atmosphere. The largest increases (six to seven degrees C.) were forecast in the interiors of northern North America and Asia during the winter; increases in the summer for the same regions were forecast at between three and four degrees C. The largest summer temperature increase (4.8 degrees C.) was forecast for interior southern Asia. The smallest increases year-round were forecast for the tropics, especially areas near large bodies of water.

An IPCC conference during November, 1990, at Geneva, Switzerland, issued a "ministerial declaration" representing 137 countries which agreed that while climate had varied in the past, "the rate of climate change predicted by the IPCC to occur over the next century [due to greenhouse warming] is unprecedented." The ministers declared, "Climate change is a global problem of unique character". The ministers also declared that the eventual goal should be "to stabilize greenhouse gas concentrations at a level that would prevent dangerous anthropogenic interference with climate".

The question of whether the Earth is becoming unnaturally warmer because of human activities was largely settled in scientific circles by 1995, with publication of the Second Assessment of the Intergovernmental Panel on Climate Change (IPCC), a worldwide group of about 2,500 experts. The panel concluded that the earth's temperature had increased between 0.5 and 1.1 degrees F. (0.3 to 0.6 degrees C.) since reliable worldwide records became available between 1850 and 1900. The IPCC noted that warming accelerated as measurements approached the present day.

The IPCC 's Second Assessment concluded that human activity increased generation of carbon dioxide and other greenhouse gases is at least partially responsible for the accelerating rise in global temperatures. The amount of carbon dioxide in the atmosphere has been rising nearly every year due to increased use of fossil fuels by ever-larger human populations experiencing higher living standards. The IPCC 's Second Assessment, according to one observer, "makes an unprecedented, though qualified, attribution of the observed climate change to human causes. Though the human signal is still building and somewhat masked within natural variation, and while there are key uncertainties to be resolved, the Panel concludes that 'the balance of evidence suggests that there is a discernible human influence on global climate'"

In its Second Assessment, the IPCC declared,

During the past few decades, two important factors regarding the relationship between humans and the Earth's climate have become apparent. First, human activities, including the burning of fossil fuels, land-use change and agriculture, are increasing the atmospheric concentrations of greenhouse gases (which tend to warm the atmosphere) and, in some regions, aerosols (microscopic airborne particles, which tend to cool the atmosphere). These changes in greenhouse gases and aerosols, taken together, are projected to change regional and global climate and climate-related parameters such as temperature, precipitation, soil moisture and sea level. Second, some human communities have become more vulnerable to hazards such as storms, floods and droughts as a result of increasing population density in sensitive areas such as river basins and coastal plains. Potentially serious changes have been identified, including an increase in some regions in the incidence of extreme high-temperature events, floods and droughts, with resultant consequences for fires, pest outbreaks, and ecosystem composition, structure and functioning, including primary productivity.

The Second Assessment indicated that rising atmospheric concentrations of greenhouse gases would cause interference with the climate system to grow in magnitude, as "the likelihood of adverse impacts from climate change that could be judged dangerous will become greater". Because of these dangers, the IPCC called upon the governments of the world to "take precautionary measures to anticipate, prevent or minimize the causes of climate change and mitigate its adverse effects".

The IPCC linked expected climate changes to rising levels of several greenhouse gases in the atmosphere following the rapid spread of fossil-fueled industry. These gases included carbon dioxide (from about 280 to almost 360 parts per million as of 1992); methane (from 700 to 1720 parts per billion); and nitrous oxide (from about 275 to about 310 parts per billion). While aerosols may have a short-term impact in some areas, the IPCC 's Second Assessment said that their short-lived nature does little to mitigate the effects of the greenhouse gases, which are resident in the atmosphere for hundreds of years.

Complexity of Atmospheric Chemistry and Dynamics

Determining the net climatic effect of a given mix of greenhouse gases is no simple matter, in part because the mix is so complex. Some gases (an example is carbon dioxide) may continue to increase in the atmosphere, while others (an example during the 1990s was methane) may stabilize or even decline. Other gases (an example being nitrous oxides) may counteract some of the effects of others.

After doubling from pre-industrial times to about 1980, the proportion of methane in the atmosphere parted paths with carbon dioxide during the 1990s, as its rate of accumulation slowed markedly, and, during 1992 and 1993, actually declined. According to a June 1999 report by E.J. Dlugokencky and associates in the British scientific journal Nature, methane levels may have declined during 1992 and 1993 for several reasons. Mount Pinatubo 's eruption in 1991 caused a cooling of global temperatures; cooler temperatures caused less methane than usual to be released from boreal wetlands.

At the same time, the countries comprising the former Soviet Union continued a decade-long decline in methane emissions from oil and natural gas production due to social and political collapse. In flusher times, the Soviet Union's oil and gas drillers vented immense amounts of methane from their operations. Decreased biomass burning in the tropics also may have contributed to the decline in atmospheric methane. The scientists who disclosed the news that methane levels have stabilized speculate that atmospheric levels will rise slowly again for a few decades, then stabilize. While scientists can measure the level of methane in the atmosphere, they have difficulty accounting for increases or decreases in that level because methane sources are many, varied, and often too small to detect in global-scale calculations.

The interactions of various greenhouse gases in the atmosphere are so complex, in some cases, that scientists are only now beginning to understand them. Mark G. Lawrence and Paul J. Crutzen surveyed the role of nitrogen oxides in the troposphere, or lower atmosphere, for example, and found that nitrous oxide "would be expected to reduce the atmospheric lifetimes of greenhouse gases such as methane as well as to increase aerosol production rates and cloud reflectivities, therefore exerting a cooling influence on the climate". According to Lawrence and his co-author, human activities account for about half the atmosphere's nitrogen oxides. Their investigation centered on nitrogen oxide emissions by ships, which eject enough of the chemicals into the air of frequently used shipping lanes to raise nitrogen oxide levels to as much as 100 times natural "background" levels.

Carbon Dioxide

The term "missing sink" with reference to the global carbon budget was coined by ocean modelers W.S. Broecker, U. Siegenthaler, and H. Oeschger during the late 1970s when they were unable to account for all the carbon released into the atmosphere from fossil-fuel combustion and land-use changes in oceanic sinks. They were uncomfortable attributing the unaccounted-for carbon to a terrestrial sink without having a detailed model to explain terrestrial biosphere carbon storage mechanisms. Scientists do not yet completely know how much of the carbon dioxide that human activity produces is stored in the ocean. There is about 60 times as much carbon dioxide in the oceans as in the atmosphere, so the interaction of the gas in the atmosphere and in the ocean is a very important element on a worldwide scale.

Of the more than seven billion tons of carbon dioxide being pumped into the atmosphere per year by human activity at the dawn of the third millennium on the Christian calendar, roughly half remained in the air. Scientists can account for about two gigatons (billion tons), which is absorbed by various carbon "sinks," mainly the oceans. A major mystery in this debate is the "missing carbon," roughly 1.5 gigatons (ggt) a year, which has evaded source detection. Plants (mainly trees) may be absorbing some of it, but with deforestation, that "sink" should be diminishing. "The mystery of the missing carbon is cause for concern," writes S. George Philander. "How can we be sure that the unidentified sink is not becoming saturated? If that should happen, the atmospheric carbon-dioxide level will soon increase more rapidly".

Some speculation has arisen that a "missing sink" of carbon exists somewhere in the terrestrial ecosystem. The size and location of such a carbon "sponge" is important to climate diplomats because, under the Kyoto Protocol, countries that harbor such carbon "sinks" could use them to balance their emissions. By 1998, some scientific argument was tending toward acceptance of a very large carbon sink over inland North America which indicates, according to Jocelyn Kaiser, that "North America may have drawn the winning card in the carbon-sink sweepstakes".

Nature may not intend irony, but its imprint here is visible to anyone who studies the politics of climate change. The proposed North American carbon sink, which is posited by S. Fan and associates, is large enough, according to Kaiser, "to suck up every ton of carbon dioxide discharged annually by fossil-fuel burning in the United States and Canada". The sink proposed by Fan and colleagues is large enough to soak up nearly a quarter of the more than six billion tons of carbon generated each year by human activity worldwide. The missing sink appears to center around the Great Lakes area. A later study by David Schimel and colleagues suggests that "processes such as regrowth on abandoned agricultural land or in forests harvested before 1980 have effects as large or larger than the direct effects of CO2 and climate". This study of net carbon storage in terrestrial ecosystems of the United States between 1895 and 1993 also suggests that the contributions of these factors may change by 100 percent year to year as a result of climate variability. While this study supports the general thrust of the Fan group, the size of the carbon sink it projects is "an order of magnitude" (e.g., one-tenth) less than that projected by Fan and associates.

No general consensus exists regarding what might be causing massive carbon absorption around the Great Lakes; some of it could be caused by new growth on formerly forested lands, or previously unexamined absorption by wetlands and soils. Carbon uptake also may be increasing because of intensive application of nitrogen fertilizers in an area that is intensely farmed. The Fan group also may have overstated its results through errors in methodology. The results also may have been skewed by the eruption of Mt. Pinatubo in 1991.

The North American carbon sink is far from a scientific done deal, however. None of the proposed causes account for its size, and even the authors of this controversial idea admit that "its magnitude remains uncertain and its cause unknown" (Fan et al. 1998, 445). Several other scientists have criticized this study's methodology, and another study indicates another, very large, carbon sink over tropical South America. This study indicates that biomass is being generated in tropical forests more quickly than it is being consumed by deforestation, thus "reducing the rate of increase in atmospheric carbon dioxide". About 40 percent of the Earth's biomass exists as tropical forests, so Phillips and associates believe that "a small perturbation in this biome could result in a significant change in the global carbon cycle".

The closer scientists look at the carbon cycle, the more reasons some find to doubt previous assumptions. For example, most models of forest response to increased greenhouse-gas levels assume that increasing temperatures will provoke the decomposition of organic matter which will raise the carbon dioxide levels in the lower atmosphere, eventually raising temperatures. This assumption is reflected in research cited in previous pages of this volume. In April,

2000, however, Christian P. Giardina and colleagues made a case in Nature that this assumption may be mistaken.

In two separate studies, new measurements of carbon flux from individual forests indicated that "decomposition of organic matter is not very sensitive to temperature" and "[soil] respiration is [an] important component of the carbon balance in northerly latitudes despite the cold temperatures there". Giardina and Ryan compiled measurements from 82 sites on five continents, while Valentini and colleagues obtained their measurements from 15 European forests. If these results are confirmed by other studies, climate modelers may be forced to revise their work. "Does this mean that the doomsday view of runaway global warming now seems unlikely? We hope so," write John Grace and Mark Rayment.

Roger M. Gifford and colleagues assert that increasing temperatures and carbon dioxide levels in the atmosphere will, up to a point, cause creation of greater plant mass, which will absorb a significant amount of the additional carbon. They contend that the ability of plants to absorb additional carbon in this manner "is now well below saturation". They add:

The modeled magnitude of the CO2 fertilizing effect on net C [carbon] storage by the biosphere is approximately the amount needed to account for the missing sink in the global C budget. Given this fact, the option must remain open that the terrestrial biosphere is responding to the increasing global atmospheric CO2 concentration, with support in some areas from the deposition of anthropogenic N [nitrogen].

The North American carbon sink is far from a scientific done deal, however. None of the proposed causes account for its size, and even the authors of this controversial idea admit that "its magnitude remains uncertain and its cause unknown". Several other scientists have criticized this study's methodology, and another study indicates another, very large, carbon sink over tropical South America. This study indicates that biomass is being generated in tropical forests more quickly than it is being consumed by deforestation, thus "reducing the rate of increase in atmospheric carbon dioxide". About 40 percent of the Earth's biomass exists as tropical forests, so Phillips and associates believe that "a small perturbation in this biome could result in a significant change in the global carbon cycle".

The closer scientists look at the carbon cycle, the more reasons some find to doubt previous assumptions. For example, most models of forest response to increased greenhouse-gas levels assume that increasing temperatures will provoke the decomposition of organic matter which will raise the carbon dioxide levels in the lower atmosphere, eventually raising temperatures. This assumption is reflected in research cited in previous pages of this volume. In April, 2000, however, Christian P. Giardina and colleagues made a case in Nature that this assumption may be mistaken.

In two separate studies, new measurements of carbon flux from individual forests indicated that "decomposition of organic matter is not very sensitive to temperature" and "soil respiration is an important component of the carbon balance in northerly latitudes despite the cold temperatures there". Giardina and Ryan compiled measurements from 82 sites on five continents, while Valentini and colleagues obtained their measurements from 15 European forests. If these results are confirmed by other studies, climate modelers may be forced to revise their work. "Does this mean that the doomsday view of runaway global warming now seems unlikely? We hope so," write John Grace and Mark Rayment.

Roger M. Gifford and colleagues assert that increasing temperatures and carbon dioxide levels in the atmosphere will, up to a point, cause creation of greater plant mass, which will absorb a significant amount of the additional carbon. They contend that the ability of plants to absorb additional carbon in this manner "is now well below saturation". They add:

The modeled magnitude of the CO2 fertilizing effect on net C [carbon] storage by the biosphere is approximately the amount needed to account for the missing sink in the global C budget. Given this fact, the option must remain open that the terrestrial biosphere is responding to the increasing global atmospheric CO2 concentration, with support in some areas from the deposition of anthropogenic N [nitrogen].

Greenhouse Gases and a Rise in Temperature

Atmospheric scientists know that the Earth's surface temperatures often have varied more or less in tandem with the atmosphere's level of carbon dioxide and other greenhouse gases.

What they do not know precisely is how much of a temperature increase may be triggered by a given rise in the level of carbon dioxide and other greenhouse gases. Atmospheric modeling is at once very simple, because the general relationship between heat-absorbing gases and temperature is known, and very complex, because climate can be influenced by many other factors as well.

Although the level of carbon dioxide in the atmosphere usually rises and falls with temperature, paleoclimatologists have found some specific epochs in the Earth's history during which glaciation occurred when carbon dioxide levels were several times higher than today, even at human-enhanced levels. One such example is the Late Ordovician (Hirnantian) glaciation during the early Paleozoic, a rare spike of cold weather during an otherwise balmy period when atmospheric carbon dioxide levels sometimes reached 14 to 16 times pre-industrial levels. Mark T. Gibbs and colleagues assert that positioning of the continents at that time may have contributed to this most unusual of ice ages.

Adding another intriguing wrinkle to the paleoclimatic puzzle, investigators have found evidence that sometimes appears to decouple carbon dioxide levels in the atmosphere from climate change. For example, M. Pagani, M.A. Arthur, and K.H. Freeman report in Paleoceanography that during the Miocene Climatic Optimum (roughly 14.5 to 17 million years ago), the Earth experienced its warmest climate in 35 million years when temperatures averaged about six degrees C. higher than today. This level of warmth was achieved with atmospheric carbon dioxide levels between 180 and 290 parts per million, compared to the present day level of about 370 p.p.m. The same researchers also found that rising, not falling, carbon dioxide levels accompanied growth in the East Antarctic Ice Sheet between 12.5 and 14 million years ago. Such findings do not necessarily contradict the infrared-forcing capacities of carbon dioxide itself but indicate that other factors (including the levels of other gases, such as methane, and effects of changes in oceanic circulation patterns) compete with carbon dioxide levels to shape the climate of any particular place on the Earth at any given time.

By the late 1990s, scientists lacking solid theoretical models which establish a direct causal link between carbon dioxide level (by itself) and temperature had largely stopped forecasting that a particular carbon dioxide level will cause a specific temperature rise. Instead, increasingly sophisticated studies are attempting to describe what role greenhouse gases will play in the context of other influences, or "forcing," in the atmosphere.

As more scientists examine the dance of climate change, they find more possible influences which complicate the identification of a specific, quantifiable role for infrared forcing on its own. For example, in the April 11, 2000 edition of the Proceedings of the National Academy of Sciences of the United States of America, Charles D. Keeling and Timothy P. Whorf propose that a 1,800-year oceanic tidal cycle is influencing climate. "We propose that such abrupt millennial changes, as seen in ice and sedimentary core records, were produced at least in part by well-characterized, almost periodic variations in the strength of the global oceanic tide-raising forces caused by resonance in the periodic motions of the Earth and Moon". Increased tidal turbulence influences the rate of "vertical mixing" in the oceans and, thus, their surface temperature. Greater oceanic mixing cools the surface of the oceans, and exerts an influence on global average temperatures. The last peak in this cycle is said, by Keeling and Whorf, to have coincided with the "Little Ice Age" which climaxed about 1600 A.D.

Impacts of Global Warming

When considering the steady rise in atmospheric carbon-dioxide levels, it is crucial (and sobering) to realize that these increases are "essentially irreversible over periods of hundreds of years". Moreover, the IPCC estimated in its Second Assessment "an immediate reduction [in emissions] of 50 to 70 percent and further reductions thereafter" to keep carbon dioxide levels in the atmosphere from rising above present levels. The IPCC presented similar scenarios for other greenhouse gases, as it warned, "The stabilization of greenhouse-gas concentrations does not imply that there will be no further climate change. After stabilization is achieved, global mean surface temperature would continue to rise for some centuries and sea level for many centuries".

By the year 2000, levels of carbon dioxide and methane in the Earth's atmosphere had risen higher than the upper range of natural concentrations during warm spells between glaciations. Until the year 2000, the "Keeling Curve" resided on uncharted atmospheric ground as far back in time as human measurement extends. By early in the year 2000, that record, from Antarctic ice cores, extended to 420,000 years before the present.

As if to illustrate how quickly baseline scientific knowledge has been evolving in this area, Paul Pearson of the University of Bristol and Martin Palmer of Imperial College, London, reported in Nature, August 17, 2000, that they have developed proxy methods for measuring the atmosphere's carbon dioxide level to 60 million years before the present. Their records considerably extend the 400,000-plus year record of ice cores. The upshot of Pearson and Palmer's studies is their conclusion that carbon dioxide levels at the year 2000 were as high as they have been in at least the last 20 million years. According to their records, however, carbon dioxide levels reached the vicinity of 2,000 p.p.m. during "the late Palaeocene and earliest Eocene periods (from about 60 to 52 million years ago)".

Pearson and Palmer used plankton shells drilled from the seabed to estimate the acidity (and thus the carbon content) of sea water over a span of time back almost to the era of the dinosaurs. By 2100, at present rates of increase, their figures indicate that the carbon dioxide level of the Earth's atmosphere may match the level last seen in the Eocene, about 50 million years ago. At that time the Earth had no permanent pack ice, and (as characterized by one English newspaper) "London was a steaming mangrove swamp".

As levels of greenhouse gases have risen, concern has been expressed by a number scientists and policy makers that the Earth may be entering a period of rapid, human-induced warming which may damage animal (including human) and plant life. The IPCC's Second Assessment indicated that deserts "are likely to become more extreme in that, with few exceptions, they are projected to become hotter but not significantly wetter". Temperature increases could threaten organisms, such as oceanic corals, which already live near their heat tolerance limits. Large areas of land may become deserts, a process which often takes the land involved to an ecological dead-end: "Land degradation in arid, semi-arid and dry sub-humid areas resulting from various factors, including climatic variations and human activities, is more likely to become irreversible if the environment becomes drier and the soil becomes further degraded through erosion and compaction".

In aquatic and coastal ecosystems, such as lakes and streams, warming would, according to the IPCC, have the greatest biological effects at high latitudes, where biological productivity would increase, and at the low-latitude boundaries of cold- and cool-water species ranges, where extinctions would be greatest. The geographical distribution of wetlands is likely to shift with changes in temperature and precipitation. Some coastal ecosystems are particularly at risk, including saltwater marshes, mangrove ecosystems, coastal wetlands, sandy beaches, coral reefs, coral atolls and river deltas.

Regional Variations of Warming Modeled

The IPCC's Second Assessment suggests that a doubling of the carbon dioxide level in the atmosphere could raise temperatures an average of 1.9 degrees C. to 5.2 degrees C. These figures are global averages which leave a great deal of room for regional variations. In general, temperatures are expected to rise most dramatically in the Arctic and Antarctic, with smaller increases nearer the equator. Temperatures also are expected to rise more in winter than in summer, and more at night than during the day.

In the North American Arctic, according to IPCC models, temperature rises may range from seven degrees C. in the lower Mackenzie Valley to more than nine degrees C. over the Arctic islands during winter, with an average summer rise of three to four degrees C. "The northward movement of permafrost could further increase the emissions of greenhouse gases into the atmosphere through the release of CO2 from soils that are highly organic, such as in Siberia," Barrie Maxwell comments. "If the soils thaw more deeply, however, methane emissions might also be reduced" because microbes will oxidize some of the methane. Maxwell notes that none of these biochemical feedbacks had been factored into existing climate models by the early 1990s.

During the late 1980s, climate researchers at England's University of East Anglia attempted to sketch conditions in a warmer world by compiling the records of a set of warm years and a set of cold years during the twentieth century. The scientists at East Anglia subjected 115 years of global temperature data to statistical analysis, and compared the result to computer climate models. They found that the correlations were far stronger for the actual temperature data than for the simulations taken from the two models. The implication of their results, the authors say, is that this century's anthropogenic greenhouse-gas-induced warming trend has overpowered natural variability.

The results of the East Anglia study corresponded rather closely to those obtained by the IPCC 's models. The greatest warming was expected in the polar regions, with a few areas, including India, the Middle East, much of Mesoamerica, and parts of Southeast Asia, expected to experience a slight cooling trend as the rest of the world warms. Precipitation may change in a more haphazard fashion, according to the East Anglia models increasing along the United States coasts, and in northern North America, Scandinavia and the Baltic States, northeastern Africa and the Middle East, India, and parts of China. Decreases are expected by this model over most of the rest of Europe and Asia, the center of North America, and Mexico.

Volcanic eruptions were found by the East Anglia models to be too infrequent to affect long-term climate significantly. The models also suggested that variations in solar output during the last century could have been large enough to influence some long-term temperature trends, however. While variations in the sun's energy output has a role in this natural equation, the East Anglia study concluded, "Solar forcing alone is insufficient to explain the behavior of the observed temperature data". The study indicated that combining solar input with changes in greenhouse gas levels produced generally credible results. Wigley said that the results "strengthen yet further our confidence that there has been a discernible human influence on climate. Furthermore, they provide additional evidence that the models used to make projections of future climate change are realistic".

Changes in landscape may modify local climates significantly. Scientists have begun to model the effects on climate of landscape changes, such as the expected northward movement of boreal forests with warming temperatures. "Because boreal forests absorb much more solar radiation than tundra does, poleward shifts in the location of the forest-tundra boundary during a period of warming can amplify climate changes by as much as fifty percent". Other studies have found that over-grazed land south of the Mexican border with Arizona experiences temperatures as much as four degrees C. warmer than nearby land on the United States side of the border which had not been grazed. The spread of irrigated farming on the eastern slope of the Rocky Mountains in Colorado has caused temperatures in some locations to fall by as much as two degrees C. Deforested areas of the Amazon Valley create "hot spots" on satellite photographs, with temperatures averaging one degree C. higher (with 30 percent less rainfall) than nearby areas which have not been logged and replaced by grassland. One observer reported in Science that some deforested areas of the Amazon resembled "a lunar landscape". In south-central Asia, irrigation has drained the Aral Sea, making summers hotter, winters colder, and the entire annual cycle drier.

Studies by R.S. Cerveny and Robert Balling support the idea that human-induced urban warming (along with increases in anthropogenic water vapor) influence downwind storminess. Cerveny and Balling examined 16 years of storm data for the eastern seaboard of the United States, and found that 20 percent more precipitation falls on weekends than week days. They contend that pollution generated in urban areas during the work week creates condensation nuclei for precipitation the following weekend.

Biotic Feedbacks: Global Warming

Human-induced warming of the Earth may be working in tandem with several natural feedback mechanisms to accelerate climate change through biotic feedbacks. The possibility that human-induced warming way feed upon itself produces a special sense of urgency in many climate scientists' public statements. Along with a sense of urgency, there is the possibility of biotic "surprises," which infuses a high degree of uncertainty into all forecasts of global warming 's possible effects.

Biotic feedbacks can enhance warming (as when deforestation decreases the Earth's ability to produce oxygen) or not (as when a jungle is replaced by a higher-albedo desert). The most important biotic feedback factors involve water vapor, the amount and nature of cloudiness, and changes in the Earth's reflectivity (albedo), most notably from high-reflection ice and snow vis-à-vis open ocean water (with, for example, a decrease of ice cover in the Arctic), which absorbs a higher proportion of incoming solar radiation. Feedbacks magnify the temperature change that would be produced by a radiative forcing, such as a carbon dioxide increase, in the absence of feedbacks.

George M. Woodwell, one of the world's most respected experts on biotic feedback mechanisms, has written, A significant body of experience suggests that there are mechanisms entrained by a change in global climate that tend to increase the trend of temperature change. There is a possibility that the warming itself may cause a series of further changes in the earth that will speed the warming. Themostserious questions have to do with the potential for surprises, especially surprises which lead to positive feedbacks. The potential appears significant.

Woodwell explains how global warming could feed powerfully upon itself:

Disruptions of forests globally, especially in the higher latitudes of the northern hemisphere, will lead to a significant increase in the release of carbon into the atmosphere. That release can easily be in the range of 1–2 billion tons per year. It means that allowing the warming to progress leads to a potential for surprises. That's only one of the surprises.

If that were to occur it would mean that correcting the problem would be even more difficult than it is at the moment. The releases from the combustion of fossil fuels at the moment are about 6 billion tons a year. The annual accumulation is 3–4 billion tons a year. Stabilizing the composition of the atmosphere would require removing from current releases something of the order of 3 billion tons, perhaps a little more, immediately. That's a half or more of the current releases of fossil fuels a very important challenge.

Warmer temperatures speed the decay of organic matter in soils. Forests and tundra soils of the Northern Hemisphere contain sufficient carbon to add significantly to the annual emissions, thereby speeding the accumulation of carbon dioxide in the atmosphere. Such a positive feedback has not been incorporated into current estimates of the warming. It is one of several potential surprises lurking in the wings as warming proceeds.

In cold water, for example, methane clath rates form crystal structures which are somewhat similar to water ice. Warming temperatures could destabilize the clathrates, and release some of their stored methane. Roughly 10 trillion tons of methane is trapped under pressure in crystal structures in permafrost or on the edges of the oceans' continental shelves, "the Earth's largest fossil-fuel reservoir," according to Gerald Dickens, a geologist at James Cook University in Townsville, Australia. The greenhouse potential of all the methane stored in clathrates on the continental shelves and in permafrost worldwide is roughly equal to that of all the world's coal reserves. Some of the land masses which host these deposits already have warmed two to four degrees C. (four to seven degrees F.) during the twentieth century.

Atmospheric scientist Roger Revelle has estimated that, with a three degree C. rise in global average temperature, methane emissions from clathrates would increase half a gigaton per year worldwide. Over a century, this rate could be enough to double the amount of methane in the atmosphere. Add to this another 12 gigatons of methane that could be released by clathrates liberated from ocean bottoms under the Arctic Ocean once the ice cap now covering them melts. "It is possible," writes Jonathan Weiner, "that this feedback effect is already underway and the rise in Earth temperatures in the last hundred years has already sprung many gigatons of methane from their molecular prisons at the bottom of the sea".

Woodwell writes,

If, for instance, a sufficient decline in the water table occurs in the boreal and tundra peatlands, subterranean fires could speed oxidation of the peat in the vast, remote peatlands of Canada and Russia, spewing forth smoke, CO2, and CH4 [methane], throughout the northern hemisphere for years. If water tables remain high, these peatlands might shift toward the production of CH4 at high rates.

Woodwell and colleagues, writing in Climatic Change, contend that while terrestrial ecosystems may have absorbed some of the increased carbon generated by human activity during most of the twentieth century, "The recent rate of increase in temperature leads to concern that we are entering a new phase in climate, one in which the enhanced greenhouse effect is emerging as the dominant influence on the temperature of the Earth".

Biotic feedbacks were discussed in Paris, during early December 1998, at a conference on climate variability organized by the World Meteorological Organization. At this conference, Stephen Schneider warned that the permafrost of Siberia and Arctic North America may already be melting and releasing methane into the air because global warming is occurring as quickly in Siberia as anywhere else on the planet.

Scientists who study biotic feedbacks sometimes remind themselves that while models are linear, nature can behave in random ways which confound linear analysis. The speed and geographic variability of ozone depletion surprised many scientists who had studied the anticipated effects of chlorofluorocarbons (CFCs) in theory. It is believed that the role of biotic feedbacks in global warming could be similarly surprising. In its Second Assessment, the IPCC stated that nonlinear systems, "when rapidly forced," are particularly subject to unexpected behavior ("surprises"). Examples of such "surprises," according to Schneider, may "include rapid decrease in the thermohaline circulation in the North Atlantic Ocean, excitation of certain dynamical modes of response of the climate system, rapid decarbonization of terrestrial ecosystems (e.g., forest die-back in fires or insect outbreaks), and catastrophic deglaciation of ice shelves in the West Antarctic".

Atmospheric levels of carbon dioxide, which increased between 1.5 and 2.4 parts per million per year during the later years of the twentieth century, build like a bank account compounding interest. Every five years, the total from which the range of increase is calculated rises by about eight percent. Add to this the fact that soils tend to release more carbon dioxide and methane naturally under warmer conditions. In scientific language, "Any changes that increase temperature or reduce pressure may liberate CH4 from hydrate. Themajor potential dangers include massive emission from Arctic hydrate, especially in western Siberia". Higher temperatures accelerate the oxidation rates of sulfur dioxide and nitrogen oxide to sulfuric and nitric acids, the precursors of acid rain.

Warming temperatures may change the behavior of the Earth's hydrological cycle. Warmer ocean water removes less carbon dioxide from the atmosphere than cooler water, so warming of the oceans may feed upon itself in coming years. Water vapor is also a potent absorber of heat in the atmosphere. It has been estimated that a doubling of carbon dioxide in the atmosphere would increase its water content by about 30 percent, raising temperatures an additional 1.4 degrees C. Many models project a rise in cloudiness, and attendant atmospheric moisture, in a warmer, more humid world. George M. Woodwell raises the possibility of a rapid surge in global warming beyond any possibility of human control:

The possibility exists that the warming will proceed to the point where biotic releases from the warming will exceed in magnitude those controlled directly by human activity. If so, the warming will be beyond control by any steps now considered reasonable. We don't know how far we are from that point because we do not know sufficient detail about the circulation of carbon dioxide among the pools of the carbon cycle. We are not going to be able to resolve those questions definitely soon. Meanwhile, the concentration of heat-trapping gases in the atmosphere rises.

Given Woodwell's expectations, the peoples of the Earth in the year 2000 are approaching a point of no return with regard to biotic feedbacks. Deforestation is accelerating around the world due to growing populations and levels of material affluence. Use of fossil fuels, which has increased at an annual rate of roughly five percent during most of this [20th] century, shows no signs of stabilizing, much less falling by half in the next 30 years. China, alone, projects burning enough fossil fuel (mainly coal) by 2025 to account for about half the present consumption of fossil fuels by everyone on Earth.

"Methane Burp" Hypothesis

Researchers who have drilled into sediment layers near the east coast of Florida found evidence that melting methane clathrates thawed suddenly (over the course of a few thousand years) about 55 million years ago, initiating a sudden episode of global warming which ended with crocodiles and palm trees in the Arctic. At the peak of this episode, greenhouse-gas levels in the atmosphere were between two and six times as high as at present. Lisa Sloan, a paleoclimatologist at the University of California (Santa Cruz), and Gerald Dickens, a paleoceanographer at James Cook University in

Queensland, Australia (two of several scientists who conducted the study), reported the findings at a meeting of the American Geophysical Union late in 1999.

The study of methane clathrates has become more popular in recent years, as evidence accumulates that their release, especially from oceans, may be a major driving force in Earth's climate cycles. James P. Kennett and colleagues studied climate records for the last 60,000 years off Santa Barbara, California, and parts of Greenland, finding that "surface and bottom temperatures change in concert". This finding supports assertions by E.G. Nisbit that massive release of oceanic methane from clathrates have played a significant role in rapid warming during the past, even without added forcing by human industry.

Scientists have yet to reach any sort of consensus on causes of the Earth's "methane burps." No one yet knows why a trillion tons of methane may be released so suddenly from solid methane hydrates around the world. This chemical reaction provoked a sudden (in geologic time) global warming of four to eight degrees C. James Cook and Gerald Dickens theorize, "The methane probably oxidized to form carbon dioxide which eventually reached the atmosphere, driving greenhouse warming".

The sediment cores drilled by Katz and associates contained remnants of small marine organisms called foraminifera, which preserve a record in their shells of carbon levels in the ocean. The shells tell a story of an extreme warming (possibly more than 10 degrees F.) in the ocean over a short time, which killed more than half of the foraminifera. The sediment core also contains evidence of an underwater landslide which scientists believe took place as melting methane clathrates "warmed dramatically, breaking apart into water and methane gas, and bubbled ferociously out of the sea floor".

This line of reasoning was supported by Richard Norris of the Woods Hole Research Center, and Ursula Rohl of Germany's University of Bremen, who wrote in Nature that the "methane burp" occurred when an as-yet-unknown natural provocation pumped greenhouse gases into the atmosphere, causing a sudden bout of global warming: "Our results suggest that large natural perturbations to the global carbon cycle have occurred in the past atrates that are similar to those induced today by human activity". Miriam E. Katz, Dorothy K. Pak, Gerald R. Dickens, and Kenneth G. Miller assert that this surge in global temperatures may have played a crucial role in the evolution of warm-blooded mammals as the Earth's dominant species 10 million years after a cataclysmic event (probably the impact, on the Earth, of a very large asteroid) ended dominance by the dinosaurs. Katz and colleagues contend that "elevated temperatures quickly opened high latitude migration routes for the widespread dispersal of mammals".

Stephen P. Hesselbo and colleagues reported that roughly 140 to 200 million years ago, large quantities of methane were liberated from ocean floors, possibly because of warming global temperatures. This "methane pulse" a "voluminous and extremely rapid release of methane from gas hydrate contained in marine continental-margin sediments" combined with oxygen in the oceans to form carbon dioxide, accelerating the worldwide warming. Along the way, a large proportion of oceanic animal life (perhaps 80 percent) died for lack of oxygen. "One of the important questions that is debated a lot today is the stability of this methane hydrate reservoir, and how easy it is to release the methane," said Stephen P. Hesselbo, lead author of the paper. "The extinction and the association with lack of oxygen has been fairly well established, but the association with methane release is something that hadn't been realized before," he said.

New Forecasts of Temperature Rise

A study issued by the Pew Center on Global Climate Change in 1999 asserts that temperatures will rise somewhat more by the year 2100 than forecast in 1995 by the Intergovernmental Panel on Climate Change's Second Assessment. The study, titled The Science of Climate Change: Global and U.S. Perspectives, was researched and written by Tom Wigley of the National Center for Atmospheric Research for the Pew Center.

The Pew Center study projects global-mean temperature increases ranging from 1.3 to 4.0 degrees C. (2.3 to 7.2 degrees F.), compared with the IPCC's projections of 0.8 to 3.5 degrees C. (1.4 to 6.3 degrees F.). The Pew Center study also forecasts a sea-level rise of 17 to 99 centimeters (7 to 39 inches) by the year 2100, compared to a previous IPCC projection of 13 to 94 centimeters (5 to 37 inches).

The study suggests that the rate of warming in the United States may be "noticeably faster than the global-mean rate" (Pew Center 1999). The study expects temperatures in the southeastern and southwestern sections of the United States to warm slightly less than the global mean. The northernmost states, from North Dakota eastward to Maine, are expected to warm as much as twice the global mean during winter months, according to Wigley 's projections. The study also forecasts, "The frequency of high-precipitation events is likely to increase, bringing increased chances of flooding". The Pew Center study also estimates that about one-third of the expected global warming during the next century may be attributed to changes in the sun's radiative output.

A report by the United Kingdom's Hadley Center for Climate Change, incorporating improved representations of ocean currents into models of the climate system, suggests that a "runaway" greenhouse effect is possible by the end of the twenty-first century. The study contends that as lack of rainfall turns large swaths of the Amazon, the Eastern United States, Southern Europe and other areas into near-deserts, the ability of plants and trees to absorb greater amounts of carbon dioxide will be reduced, resulting in higher atmospheric concentrations and rapid global temperature increases. This report projects that agricultural output in central and southern Africa will be severely reduced, and North America's agricultural heartland could see wheat and corn yields fall by as much as 10 percent. Extreme water shortages will affect 170 million people, according to this study, which was presented at the Buenos Aires climate conference in 1998. The study forecasts that temperatures on land will rise an average of 6 degrees C. by the year 2100, subjecting about 100 million people to annual hazards of coastal flooding from rising sea levels.

The Hadley Center study also projects that the Gulf Stream, which is an important warming influence on much of Europe during the winter, will be 20 percent less strong in the future, but that Europe still will warm considerably. This study, unlike some others, does not foresee a weakening of the Gulf Stream as portending colder winters for Europe while most of the rest of the world warms. Instead, the Hadley Center model forecasts that Western Europe, including Scotland, will gain the ability to grow extra grain and that European storms will become more severe, especially during the winter.

The Hadley Center study anticipates that the benefits for plants of a carbon-enhanced atmosphere will be outweighed by lack of rainfall in many important agricultural areas. The study also asserts that many tropical grasslands will be transformed into deserts, leading to widespread extinction of wildlife. Michael Meacher, British environment minister, told The Guardian that "These are sobering findings. Millions of people will have life made miserable by climate change, with increased risk of hunger, water shortages and extreme events like flooding. Combating climate change is the greatest challenge of human history". The Hadley Center study also anticipates that much of central and southern Africa will experience a reduced ability to grow staple crops. While the agricultural heartland of the United States may suffer production reductions because of drought and heat, the study projects that Canada will experience a wheat production increase of about 2.5 percent.

Temperature readings during the late 1990s indicated that a steep rise in temperatures seemed to be underway. On March 9, 2000 the National Oceanic and Atmospheric Administration (NOAA) said the winter of 1999–2000 was the warmest such season in the United States since the government began keeping records 105 years earlier. This marked the third year in a row that record warmth was recorded in the United States during the winter months. Since 1980, more than two-thirds of U.S. winters have been warmer than average, NOAA said.

The average temperature in the United States between December, 1999 and February, 2000 was 38.4 degrees F., six-tenths of a degree warmer than the record set the previous year. Scientists at NOAA reported that every state in the continental United States was warmer than its long-term average, with 21 states from California to the Midwest ranked as much above average. As the report was being released, a winter carnival was being canceled near Wausau, Wisconsin, for lack of snow. Other casualties of the warmth included North America's largest cross-county ski race and an ice fishing derby in International Falls, Minnesota.

A week later, the Great Lakes, the world's largest body of fresh water, were measured at their lowest level in recorded history, because of scant snowfall during the winter, after several years of declining water levels. Consequences have included dry wells, landlocked docks, obstacle courses for commercial shipping and pleasure boaters, and smelly drinking water in some areas. Many docks have become useless, while emergency dredging has been required for others. At the same time, the temperature of the lakes' water has been climbing.

During 2000, Buffalo, New York reported that its harbor's water temperature at the end of March had equaled the record warmest (39 degrees F.) set in 1998. During 1990, Congress commissioned a study of how global warming would affect various regions by the year 2100. The report was issued in draft for public comment during the summer of 2000. The report, which involved 5,000 people in nine federal agencies, projected that average temperatures will rise 5 to 10 degrees F. by the end of the century. This report analyzed possible climate changes in eight regions of the United States, "based on a pair of state-of-the-art climate models one from the Canadian Climate Center and one from the United Kingdom Hadley Centre for Climate Research and Prediction". The entire study was coordinated by Thomas Karl, director of the NOAA National Climate Center in Asheville, North Carolina, who said that the report illustrated a "range of our uncertainties." To Karl, the report also indicated that "The past isn't going to be a very good guide to future climate".

The two climate models used in the study sometimes contrast sharply. The Canadian model, for example, indicates frequent severe drought in the United States' agricultural heartland, while the Hadley model suggests plentiful rainfall in the same area. Generally, however, the report supports a 5 to 10 degree F. rise in temperatures during the century, in line with the models of the Intergovernmental Panel on Climate Change. The report was largely a compilation of data from existing sources because when Congress mandated the study no funds were provided to pay for it.

According to this report, water levels in the Great Lakes are expected to drop five feet by century's end. By the year 2000, Lakes Erie, Michigan, and Huron had dropped three feet in three years, while Lakes Superior and Ontario were down about 18 inches during the same period. Todd Thompson of the Indiana Geological Survey was quoted as saying that the levels of the Great Lakes ebb and flow in 30-year cycles. The lakes receded during the drought years of the 1930s, then again in the 1960s. Coming years will reveal whether present lake-level declines are merely cyclical, or part of a new trend related to global warming. In the meantime, the Toledo Beach Marina was spending $1 million during the year 2000 to add three and a half feet of draft to its docking facilities. Lake Erie averages only 70 feet deep, and in some places a few inches makes a big difference for cargo shipping. At about the same time, the National Environmental Trust released a report describing global warming 's anticipated effects on the Great Lakes. Philip Clapp, president of the organization, said dredging of shipping lanes caused by declining lake levels could cost billions of dollars.

The same federal report also anticipates some beneficial effects of warming, including reduced costs for snow removal in many mid western cities, and an opportunity for farmers to profit by planting more than one crop a year.

As news of recent temperature rises arrived during the year 2000, the IPCC strengthened its previous statement affirming human modification of the Earth's climate. In 1995 the IPCC had concluded that "the balance of evidence suggests a discernible human influence." In its 2000 assessment, this language reads: "There has been a discernible human influence on global climate". The temperature records of the last 1,000 years now leave very little doubt that the upward spike in temperatures since 1980 has been influenced in large part by human greenhouse-gas emissions. The IPCC 's new forecasts for global temperatures in 2100 changed little in 2000 from the 1990 or 1995 assessments one degree C. to five degrees C., according to several sets of assumptions about how human numbers, societies, economies, and technologies may change during that time.

Possible Speed of Climate Change

Twenty-one nationally prominent ecologists warned President Clinton that rapid climate change due to global warming could ruin ecosystems on which human societies depend. The signers, including Stephen H. Schneider and three colleagues from Stanford University, urged Clinton to take a "prudent course" in the then-upcoming global climate-change negotiations in Kyoto, Japan. The scientists warned that the warming would happen so quickly that many plant and animal species will not be able to adapt. The resulting breakdown of ecosystems could lead to disturbances with major effects on human populations, the scientists warned. These may include increasing numbers of fires, floods, droughts, and storms, as well as erosion and outbreaks of pests and pathogens. The letter said that if present levels of greenhouse-gas emissions continue to rise, the climate will change more quickly during the coming century than at any time during the past 10,000 years.

"The signers include the leading international experts on many particular dimensions of this problem," said Harold Mooney, Stanford professor of biological sciences and the organizer of the effort. "As you will read in the letter, they all have deep concerns about the ecological consequences of rapid climatic change". Among the signers are Mooney, as well as Paul Ehrlich of Stanford (an international leader in ecological research), and Jane Lubchenco of Oregon State University, a past president of the American Association for the Advancement of Science. Seven of the signers are members of the National Academy of Sciences and five are past presidents of the Ecological Society of America.

The scientists said, in the United States

Rapid climate change could mean the widespread death of trees, followed by wildfires and replacement of forests by grasslands. National parks and forests could become inhospitable to the rare plants and animals that are preserved there and where the parks are close to developed or agricultural land, the species themselves may disappear for lack of another safe haven. Worldwide, fast-rising sea levels could inundate the marshes and mangrove forests that protect coastlines from erosion and serve as filters for pollutants and nurseries for ocean fisheries. "The more rapid the rate [of change] the more vulnerable to damage ecosystems will be," the scientists told the president. "We are performing a global experiment with little information to guide us.

The ecologists warned that in some United States temperate-zone forests, rapid climate change could lead to "widespread tree mortality, wildfires and replacement of the forests by grasslands. Species that are long-lived, rare, or endangered will be severely disadvantaged". "It would be difficult to imagine, for example," the scientists wrote, "how the imperiled species of Everglades National Park, such as the Cape Sable Sparrow and American Crocodile, could migrate north into the urban and agricultural landscapes of coastal and central Florida and successfully re-establish themselves".

The scientists' letter seemed to have an impact at the White House, judging from presidential rhetoric. In his 1999 State of the Union speech, President Clinton called global warming "our most fateful new challenge," as he recalled 1998 as the warmest year ever recorded, with heat waves, floods, and storms which "are but a hint of what future generations may endure if we do not act now." Clinton proposed creation of a new "clean air fund to help communities reduce greenhouse and other pollutions." Clinton also said he "want[ed] to work with members of Congress in both parties to reward companies that take early, voluntary action to reduce greenhouse gases". Many of Clinton's proposals were repeated a year later in his 2000 (and last) State of the Union speech.

"We know from ice-core records and deep-sea sediment records that the earth's climate is capable of changing much more quickly than we had previously thought," said Jeff Severinghaus of the University of California. "In some cases," said Severinghaus, "the climate warmed abruptly in less than 10 years up to possibly 10 degrees centigrade". Severinghaus 'findings were presented at the 1998 climate-change conference in Buenos Aires.

Severinghaus continued,

It is possible that by increasing greenhouse gases, we will induce such a change and that, instead of the smooth warming that's being anticipated over the next 50 years, we'll instead go along for a while with very little warming and then all of a sudden in a matter of three or five or ten years we'll have a very large catastrophic warming.

At the National Ice Core Laboratory in Denver, thousands of meter-long tubes are arrayed on shelves, holding ice cores from the Arctic and Antarctic at minus 36 degrees C. These ice cores contain records of the Earth's changing climate for the last 420,000 years. From studies of ice cores taken in Greenland, scientists have assembled a climatic record which indicates that during the last 8,000 years the earth's climate has been relatively mild and stable. At the end of ice ages (the most recent one, which ended about 12,000 years ago is an example), temperatures tend to swing wildly in cycles of five to twenty years. According to Gale E. Christianson,

"Temperatures rose by an astonishing 10 degrees C. within the lifespan of a Paleolithic hunter, and some scientists now think that even that figure is too low by half".

Richard B. Alley, also writing in the Proceedings of the National Academy of Sciences, reviewed ice-core evidence from the last 110,000 years which indicates that climate may vary very little over long periods, then undergo changes as large as those between glacial and interglacial conditions, sometimes within a few years or decades. Alley points out that the development of complex human civilization has taken place during a period without such rapid changes.

Thomas V. Lowell of the University of Cincinnati's Geology Department, also writing in the Proceedings of the National Academy of Sciences, used changes in glacial mass to track climatic change. Using such measures, Lowell provides graphic evidence (from glacial samples near coastal Alaska's Prince William Sound and Western Greenland) that average temperatures declined slowly from about 1450 A.D. to almost 1900 A.D. (the so-called "Little Ice Age"). At that time, temperatures began to climb rapidly, at a pace of about 0.8 degrees C. per century, four times the rate of change during the previous 900 years. The temperature curve rises at an increasingly steep angle toward the end of the century.

Jonathan Overpeck, director of the paleoclimatology program for the National Oceanic and Atmospheric Administration (NOAA), said at the end of 1998, "There is no period that we can recognize in the last 1,200 years that was as warm on a global basis [as the present]." Overpeck presented his findings at a meeting of the American Geophysical Union in San Francisco. "That makes what we're now seeing more unusual, and more difficult to explain without turning to a 'greenhouse gas' mechanism," said Overpeck. By the 1990s, according to the IPCC 's Second Assessment the temperature was rising at the most rapid rate in at least 10,000 years.

Until the 1990s, many climate scientists believed that the Earth had warmed dramatically during the period of time which Europe called the Middle Ages, roughly 900 to 1400 A.D. New research, based on tree rings, glaciers and other "proxy" measurements of past climate around the world, indicates that this warming was limited mainly to northern latitudes in Europe and North America. Evidence of a rapid warm-up during the Middle Ages has been used as "proof" by some global-warming skeptics that natural variations may explain rapid temperature increases worldwide during the last quarter of the twentieth century. "Our study of the Medieval Warm Period supports the likelihood that no known natural phenomenon can explain the record twentieth-century warmth," Overpeck said. "Twentieth century global warming is a reality and should be taken seriously".

Dean Edwin Abrahamson confirms Overpeck 's analysis:

One must go back in time 5 to 15 million years to the late Tertiary to find a time that was 3 or 4 degrees C. warmer than now. During periods when there was no permanent pack ice in the Arctic, climatic and vegetational region and boundaries were displaced as much as 1,000 to 2,000 kilometers north of their present position (a displacement which we may replicate during the next 100 years).

During the period Abrahamson describes, intense aridity was the norm from present day North and South Dakota to Missouri and Alabama, as well as throughout Central and Southern Africa. These changes may be similar to those which will be experienced by generations to come. As Abrahamson comments,

We could be committed to a far larger warming probably on the order of 6 to 10 degrees C. The climatic conditions that might be associated with such a warming are, with few exceptions, pure mystery. Today's climate models have little, if any, validity for such extreme warming. There can be no planned adaptation under these conditions.

By early in the year 2000, scientists working for NOAA released compilations of global temperatures for the last half of the twentieth century which reveal a speed of warming that most climatologists had not expected until late in the twenty-first century. The rate of warming (one degree F. over the entire century) increased to a rate of four degrees F. during the century's last quarter, according to calculations of Tom Karl and associates, published in the March 1, 2000 edition of Geophysical Research Letters. This is roughly the rate of increase which several climate models had forecast for the second half of the twenty-first century. "The next few years could be very interesting," Karl told the Los Angeles Times. "It could be the beginning of a new increase in temperatures". Tom Wigley, a senior scientist at the National Center for Atmospheric Research in Boulder, Colorado, said that warming was strengthened by frequent El Nino events, which he said are not human-induced. "Those months were unusual," he said, "but they weren't unusual due to human influences".

Karl and colleagues begin a statistical analysis of recent global temperature trends with the observation that between May of 1997 and September of 1998, sixteen months in a row, global temperatures set observational (e.g., century-scale) monthly records. Their analysis of a century-plus of records (roughly 1880 to 2000) indicates that the rate of warming tends to surge upward, then relent a little, and then surge again. "The increase in global mean temperatures is by no means constant". Karl and colleagues conclude, "The warming rate over the past few decades [since the mid-1970s] is already comparable to that projected during the twenty-first century based on IPCC business-as-usual scenarios of anthropogenic climate change".

We interpret the results to indicate that the mean rate of warming since 1976 is clearly greater than the mean rate of warming averaged over the late nineteenth and twentieth centuries. It is less certain whether the rate of temperature change has been constant since 1976 or whether the recent string of record-breaking temperatures represents yet another increase in the rate of temperature change. Moreover, these results imply that if the climate continues to warm at present rates of change, more events like the 1997 and 1998 record warmth can be expected.

Ozone Depletion's Relationship

Chlorofluorocarbons (CFCs) initially raised no environmental questions when they were first marketed by Dupont Chemical during the 1930s under the trade name Freon. Freon was introduced at a time when such questions usually were not asked. At about the same time, asbestos was being proposed as a high-fashion material for clothing, and radioactive radium was being built into time-pieces so that they would glow in the dark.

Manufacturers in the United States were producing 750 million pounds of CFCs a year, and finding all sorts of uses for them, from propellants in aerosol sprays, to solvents used to clean silicon chips, to automobile air conditioning, and as blowing agents for polystyrene cups, egg cartons, and containers for fast food. "They were amazingly useful," wrote Anita Gordon and Peter Suzuki. "Cheap to manufacture, non-toxic, non-inflammable, and chemically stable". By the time scientists discovered, during the 1980s, that CFCs were thinning the ozone layer over the Antarctic, they found themselves taking on a $28 billion-a-year industry. The ozone shield protects plant and animal life on land from the sun's ultraviolet rays, which can cause skin cancer, cataracts, and damage to the immune systems of human beings and other animals. Thinning of the ozone layer also may alter the DNA of plants and animals.

By the time their manufacture was banned internationally during the late 1980s, CFCs had been used in roughly 90 million car and truck air conditioners, 100 million refrigerators, 30 million freezers, and 45 million air conditioners in homes and other buildings. Because CFCs remain in the stratosphere for up to 100 years, they will deplete ozone long after industrial production of the chemicals ceased.

These human-created chemicals do more than destroy stratospheric ozone. They also act as greenhouse gases, with several thousand times the per-molecule greenhouse potential of carbon dioxide. What's more, the warming of the near-surface atmosphere (the lower troposphere) seems to be related to the cooling of the stratosphere, which accelerates depletion of ozone at that level. An increasing level of carbon dioxide near the Earth's surface "acts as a blanket," said NASA research scientist Katja Drdla. "It is trapping the heat. If the heat stays near the surface, it is not getting up to these higher levels".

At about the same time, scientists were looking for reasons why the ozone layers over the Arctic and Antarctic were failing to repair themselves as expected following the international ban on production of CFCs. They began to suspect that global warming near the surface might be related to ozone depletion in the stratosphere.

During the middle 1990s, scientists were beginning to model a relationship between global warming and ozone depletion. A team led by Drew Shindell at the Goddard Institute for Space Studies created the first atmospheric simulation to include ozone chemistry. The team found that the greenhouse effect was responsible not only for heating the lower atmosphere, but also for cooling the upper atmosphere. The cooling poses problems for ozone molecules, which are most unstable at low temperatures. Based on the team's model, the buildup of greenhouse gases could chill the high atmosphere near the poles by as much as 8 degrees C. to 10 degrees C. The model predicted that maximum ozone loss would occur between the years 2010 and 2019.

In 1998, the Antarctic ozone hole reached a new record size roughly the size of the continental United States. Some researchers came to the conclusion that, as Richard A. Kerr describes in Science,

Unprecedented stratospheric cold is driving the extreme ozone destruction. Some of the high-altitude chill may be a counterintuitive effect of the accumulating greenhouse gases that seem to be warming the lower atmosphere. The colder the stratosphere, the greater the destruction of ozone by CFCs.

"The chemical reactions responsible for stratospheric ozone depletion are extremely sensitive to temperature," Shindell and colleagues wrote in Nature. "Greenhouse gases warm the Earth's surface but cool the stratosphere radiatively, and therefore affect ozone depletion". By the decade 2010 to 2019, Shindell and colleagues expect ozone loses in the Arctic to peak at two-thirds of the "ozone column," or roughly the same ozone loss observed in Antarctica during the early 1990s. "The severity and duration of the Antarctic ozone hole are also expected to increase because of greenhouse-gas-induced stratospheric cooling over the coming decades".

During the middle 1990s, scientists began to detect ozone depletion in the Arctic after a decade of measuring a growing ozone "hole" over the Antarctic. By the year 2000 during March and April, the ozone shield over the Arctic had thinned to about half its previous density. Ozone depletion over the Arctic reaches its height in late winter and early spring, as the sun rises after the midwinter night. Solar radiation triggers reactions between ozone in the stratosphere and chemicals containing chlorine or bromine. These chemical reactions occur most quickly on the surface of ice particles in clouds, at temperatures less than minus 80 degrees C. (minus 107 degrees F.).

Space-based temperature measurements of the Earth's lower stratosphere, a layer of the atmosphere from about 17 kilometers to

22 kilometers (roughly 10 to 14 miles) above the surface, indicate record cold at that level as record surface warmth has been reported during the 1990s. Roy Spencer of NASA and John Christy of the University of Alabama at Huntsville and the Global Hydrology and Climate Center obtained temperature measurements of layers within the entire atmosphere of the Earth from space, using microwave sensors aboard several polar-orbiting weather satellites. They found that, despite significant, short-lived warming following the eruptions of El Chichon in Mexico in 1982 and Mt. Pinatubo in the Philippines in 1991, the stratosphere as a whole has been cooling steadily during the past 15 years.

Steve Hipskind, atmospheric and chemistry dynamics branch chief at NASA's Ames Research Center, Moffett Field, California, has been quoted as saying that chlorine atoms use clouds as "a platform" to destroy stratospheric ozone. Clouds form more frequently over the Arctic at lower temperatures. Ice crystals, which form as part of polar stratospheric clouds, assist the chemical process by which ozone is destroyed. CFCs' appetite for ozone molecules rises notably below minus 80 degrees C. (minus 107 degrees F.), a level that was reached in the Arctic only rarely until the 1990s. During the winter of 1999–2000, temperatures in the stratosphere over the Arctic were recorded at minus 118 degrees F. or lower (the lowest on record), forming the necessary clouds to allow accelerated ozone depletion.

As Dennis L. Hartmann and colleagues explain,

The pattern of climate trends during the past few decades is marked by rapid cooling and ozone depletion in the polar lower stratosphere of both hemispheres, coupled with an increasing strength of the wintertime westerly polar vortex and a poleward shift of the westerly wind belt at the Earth's surface. Internal dynamical feedbacks within the climate system can show a large response to rather modest external forcing. Strong synergistic interactions between stratospheric ozone depletion and greenhouse warming are possible. These interactions may be responsible for the pronounced changes in tropospheric and stratospheric climate observed during the past few decades. If these trends continue, they could have important implications for the climate of the twenty-first century.

Ozone depletion has been measured only for a few decades, so these researchers caution that they are not entirely certain that rapid warming at the surface is not caused by natural variations in climate, which is powerfully influenced by the interactions of oceans and atmosphere. "However," they conclude, "it seems quite likely that they are at least in part human-induced." Hartmann and associates also raise the possibility that the poleward shift in westerly winds may be accelerating melting of the Arctic ice cap, part of what they contend may be a "transition of the Arctic Ocean to an ice-free state during the twenty-first century." A continued northward shift in these winds also could portend additional warming over the land masses of North America and Eurasia, they write.

The connection between global warming, a cooling stratosphere, and depletion of stratospheric ozone was confirmed in April, 2000, with release of a lengthy report by more than 300 NASA researchers as well as several European, Japanese, and Canadian scientists. The report found that while ozone depletion may have stabilized over the Antarctic, ozone levels north of the Arctic circle were still falling, in large part because the stratosphere has cooled as the troposphere has warmed. The ozone level over parts of the Arctic was 60 percent lower during the winter of 2000 than during the winter of 1999, measured year over year.

In addition, scientists learned that as winter ends, the ozone-depleted atmosphere tends to migrate southward over heavily populated areas of North America and Eurasia. "The largest ultraviolet increases from all of this are predicted to be in the mid-latitudes of the United States," said University of Colorado atmospheric scientist Brian Toon. "It affects us much more than the Antarctic [ozone "hole"]".

Ross Salawitch, a research scientist at NASA's Jet Propulsion Laboratory in Pasadena, California, said that if the pattern of extended cold temperatures in the Arctic stratosphere continues, ozone loss over the region could become "pretty disastrous". Salawitch said that the new data has "really solidified our view" that the ozone layer is sensitive not only to ozone-destroying chemicals, but also to temperature. "The temperature of the stratosphere is controlled by the weather that will come up from the lower atmosphere," said Paul Newman, another scientist who took part in the Arctic ozone project. "If we have a very active stratosphere we tend to have warm years, when stratosphere weather is quiescent we have cold years". New research indicates that global warming will continue to cool the stratosphere, making ozone destruction more prevalent even as the volume of CFCs in the stratosphere is slowly reduced. "One year does not prove a case," said Paul Newman of NASA's Goddard Space Flight Center in Greenbelt, Maryland. "But we have seen quite a few years lately in which the stratosphere has been colder than normal".

"We do know that if the temperatures in the stratosphere are lower, more clouds will form and persist, and these conditions will lead to more ozone loss," said Michelle Santee, an atmospheric scientist at NASA's Jet Propulsion Laboratory in Pasadena and co-author of a study on the subject in the May 26, 2000 issue of Science. The anticipated increase in cloudiness over the Arctic could itself become a factor in ozone depletion. The clouds, formed from condensed nitric acid and water, tend to increase snowfall, which accelerates depletion of stratospheric nitrogen. The nitrogen (which would have acted to stem some of the ozone loss had it remained in the stratosphere) is carried to the surface as snow.

2

Icemelt and Glacial for Global Warming

Introduction

The rising oceans would inundate intensely farmed river deltas, such as the Nile and Ganges. In the real world, however, the melting of the West Antarctic Ice Sheet may take several centuries to unfold, and thus will not reach its climatic end until long after the last members of any present day audience will have left the theater. Ice turns to water one prosaic drop at a time.

Even though the melting of the world's ice usually lacks Hollywood-style flair, the eventual result may be climate change on a scale that the Earth has not experienced since the days of the dinosaurs. Human-induced climate change may be ending the cycle of glaciation that has been the norm on Earth since the end of the age of the dinosaurs, about 65 million years ago. Viewed on a timescale of the last 540 million years, however, glacial cycles have been relatively rare events. For three-quarters of the last 540 million years, the Earth's ice caps have been negligible or non-existent.

As the twenty-first century opened, ice was melting around the world. The Arctic ice cap was thinning, and scientists were speculating over how many years will pass before a summer during which most of it melts. The Intergovernmental Panel on Climate Change's models project that between one-third and one-half of existing mountain glacial mass could disappear over the next 100 years. Sometime during the present century, the last glacier may melt in Glacier National Park.

Many lakes and rivers in the Northern Hemisphere usually freeze about a week later and thaw out 10 days sooner than a century and a half ago, according to John J. Magnuson and colleagues. In some areas (such as the harbor of Toronto, Ontario), heat contributed by urbanization, as well as general warming, may have added a month or more to the ice-free season. The warming trend appears to have begun at least half a century before the significant buildup of greenhouse gases caused by the burning of fossil fuels, leading to speculation that a long-term climate cycle is accelerating changes induced by infrared forcing.

An international team of scientists analyzed written accounts, some of them centuries old, including newspaper reports, fur traders' records, ships' navigation logs, and descriptions of religious events. These results illustrate "a very clear record of the response of aquatic systems to global warming," said Magnuson. The authors of this study estimate that the changes in freezing and thawing dates that they found required an average temperature increase of about three degrees F. during the last 150 years. "We know that the regions of Eurasia and North America, where most of these [lake and river sites] are, has indeed warmed according to thermometers at a rate of about double that for the globe both recently and for the past century," said Kevin Trenberth, head of climate analysis at the National Center for Atmospheric Research, who did not participate in the research.

Observed ice melt probably accounts for only a fraction of the impact to come, as the world's oceans and atmosphere factor increased levels of "greenhouse forcing" into their dynamics. Jonathan Overpeck, a paleoclimate specialist with the National Oceanic and Atmospheric Administration (NOAA) in Boulder, Colorado, speculates that human activity may be short circuiting the glaciation cycle which has dominated the Earth's climate for millions of years.

If we left Mother Nature to do what she wanted to do, we would be going back into another ice age in the next 10,000 years. Now, because of what humans are doing, it's unlikely that we'll be going back into another ice age. Instead, glaciers around the world are receding. The Earth is warming up, and it will likely continue to warm up.

As the Earth warms, increased advection transports surplus heat from equatorial to polar areas (heat flows from hotter to colder regions), so temperature changes tend to be larger at the poles than at mid-latitudes. In the tropics, warming generally has been more pronounced at higher altitudes than closer to sea level.

Harvard epidemiologist Paul Epstein comments, Some of the strongest evidence of a warming trend comes from mountainous regions, where summit glaciers are melting on six continents. Many may soon disappear. In tandem with the retreat of peak ice sheets, plants are migrating up mountain slopes on 26 Swiss Alpine crests, and in the United States Sierra Nevada, Alaska and New Zealand.

The melting of Arctic and Antarctic ice will do more than inconvenience coastal urban dwellers. Melting is already destroying an ecosystem built around sea ice. A report by the World Wildlife Fund and the Marine Conservation Biology Institute sketches the vital role of sea ice in polar ecosystems:

Sea ice is fundamental to polar ecosystems: it provides a platform for many marine mammals and penguins to hunt, escape predators, and breed. It sedges and undersides provide vital surfaces for the growth of algae that forms the base of the polar food web. In areas with seasonal ice cover, spring blooms of phytoplankton occur at ice edges as the ice cover melts, boosting productivity early in the season. But sea ice is diminishing in both the Arctic and the Antarctic. As this area diminishes, so does the food available to each higher level on the web, from zooplankton to seabirds. Higher temperatures predicted under climate change will further diminish ice cover, with open water occurring in areas previously covered by ice, thereby diminishing the very basis of the polar food web.

Reduced ice cover changes the Earth's albedo (reflectivity), which could become a factor in global warming, causing the earth to absorb more solar energy, as ice and snow (which reflect 75 percent or more of incoming sunlight) is replaced by bare soil, which reflects 10 to 25 percent. Ice and snow, in some cases, may be replaced by liquid seawater, which reflects 10 to 70 percent of incoming sunlight, depending on the sun's angle.

West Antarctic Ice Sheet

The idea that global warming could provoke the disintegration of the West Antarctic Ice Sheet was aired as theory by glaciologists as early as 1979. J.H. Mercer has suggested that the West Antarctic Ice Sheet fell apart during an interglacial period about 125,000 years ago without an added boost from the burning of carbon-based fuels. T.J. Hughes has examined the geophysical mechanisms which may cause the West Antarctic Ice Sheet to collapse, and J. T. Hollin has examined evidence of major ice-sheet "surges" in the past which led to 10 to 30 meter rises of sea level in less than 100 years.

During the 1990s, the stability of the West Antarctic Ice Sheet (which comprises about a quarter of the Earth's largest mass of frozen water) became a subject of intense scientific inquiry. A vibrant debate has grown up regarding the future of the ice sheet, with assurances of stability on one side, and speculation of future collapse on the other.

A report issued by the National Academy of Sciences (NAS) during 1991 asserted that melting of the West Antarctic Ice Sheet is unlikely, "and virtually impossible before the end of the next century". According to climate models used in this report, several centuries of rising temperatures will be required before the ice sheet disintegrates.

Reports authored by committees sometimes illustrate disagreements between the lines. One page after it dismisses the prospect that the West Antarctic Ice Sheet may melt, the same NAS report says that existing climate models may not allow for unforeseen "surprises." The literature of global warming is studded with admissions of this kind, which lend a startling degree of uncertainty to any attempt to forecast future climatic conditions and their impact around the world. For example, the same NAS report says, "CH4 [methane] could be released as high-latitude tundra melts, providing a sudden increase in CH4, which would add to the greenhouse warming" and "There could be a significant melting of the West Antarctic Ice Sheet resulting in a sea-level several meters higher than it is today".

During 1998, an international team of scientists analyzed five years of satellite radar measurements covering a large part of the West Antarctic Ice Sheet to determine whether it is becoming more unstable. The team concluded that the ice sheet is not melting rapidly and remains reasonably stable, as it has been for more than a century. "Based on our short, five-year period of observation of the interior of Antarctica, we do not seem to detect that the ice is melting more than one centimeter per year," explained C.K. Shum, an associate professor of civil and environmental engineering at Ohio State University. "That would mean that the interior Antarctic ice sheet does not seem to be contributing to sea-level rise more than 1 millimeter per year. We assume that global warming is under way now, and it may be enhanced by human activities but, until now, its effect on ice loss in Greenland and the Antarctic has been mostly speculation".

The rate of ice decay on the Antarctic Peninsula accelerated markedly during the record global warmth of the 1990s. A 48-by-22-mile chunk of the Larsen Ice Shelf broke off during March, 1994, exposing rocks that had been buried for 20,000 years, prompting Rodolfö del Valle, director of geoscience at the Argentine Antarctic Institute in Buenos Aires, to tell the Associated Press, "Last November we predicted the ice shelf would crack in ten years, but it has happened in barely two months". Del Valle was one of a team of scientists who witnessed the collapse of the Larsen A Ice Shelf, a floating ice sheet as large as Rhode Island which had averaged 500 feet in thickness. One day, according to Christian Science Monitor correspondent Colin Woodard, the shelf collapsed with a thunderous roar. In two hours, the Argentine team found itself standing on an island surrounded by open water and enormous icebergs. "I felt a sadness, a pain in my heart for the loss of a place that had become like a home to me," del Valle recalls. "I've experienced strong earthquakes on land, but this was different. After an earthquake something remains. But not with the ice shelf it was completely destroyed".

In early 1995, the Larsen A Ice Shelf completely disintegrated during a single storm after years of shrinking gradually. "The speed of the final breakup was unprecedented, and followed several of the warmest summers on record for this portion of the Antarctic," said Ted Scambos of the National Snow and Ice Data Center based at the University of Colorado at Boulder. "Ice shelves appear to be good bellwethers for climate change, since they respond to change within decades, rather than the years or centuries sometimes typical of other climate systems," said Scambos.

During 1998 and 1999, several reports described the retreat of ice along the shores of the Antarctic Peninsula. Reports indicated that an additional 1,100 square miles of ice had melted during 1998. At about the same time, David Vaughan, a researcher with British Antarctic Survey, and Scambos reported that the Larsen B and Wilkins shelves on the Antarctic Peninsula were in "full retreat". These ice sheets had been retreating slowly for about 50 years, losing about 2,700 square miles during that period. A loss of 1,100 square miles in one year thus represented a major acceleration of the ice sheets' erosion. During that year as much ice was lost as had melted during the preceding four decades.

"This may be the beginning of the end for the Larsen Ice Shelf," said Scambos as the ice sheet crumbled in 1999. "This is the biggest ice shelf yet to be threatened," Scambos said. "The total size of the Larsen B Ice Shelf is more than all the previous ice that has been lost from Antarctic ice sheets in the past two decades". Until its dissolution, the Larsen B was the northernmost ice shelf in Antarctica, and therefore "on the front line of the warming trend," said Scambos.

Radar images from satellite observations of the Pine Island glacier in Antarctica taken during the 1990s indicate that it has been shrinking rapidly. The shrinking of this glacier is important "because it could lead to a collapse of the West Antarctic Ice Sheet," said Eric Rignot, a computer-radar scientist at the Jet Propulsion Laboratory in California, who led a study of the glacier. "We are seeing a glacier melt in the heart of Antarctica". "The continuing retreat of Pine Island glacier could be a symptom of the WAIS [West Antarctic Ice Sheet] disintegration," said Craig Lingle, a glaciologist at the University of Alaska in Fairbanks, who is familiar with the study.

The Pine Island glacier is important because it is part of a stream of ice that moves more rapidly than the ice cap surrounding it. This glacier is part of an ice stream which runs from the interior of the West Antarctic Ice Sheet into the surrounding ocean waters. If a glacier in this ice stream melts more quickly from the bottom than snow accumulates on its top, the net icemelt goes into the ocean, raising sea levels.

A "disaster scenario," as described by Richard Alley, a glaciologist at Pennsylvania State University, has the Pine Island glacier retreating enough to "make a hole in the side of the ice sheet. The remaining ice would drain through that hole". Once enough ice had drained through the hole, the West Antarctic Ice Sheet might eventually collapse, raising average sea levels around the world 15 to 20 feet in a few years. Such an increase in mean sea level would flood roughly 30 percent of Florida and Louisiana, 15 to 20 percent of the District of Columbia, Maryland, and Delaware, and 8 to 10 percent of the Carolinas and New Jersey. Inundation of coastal areas would have a similar impact around the world. Among the flooded areas would be the centers of some of the world's great urban and commercial centers, from New York City to Bombay, Calcutta, and Manila.

During 1998, Eric Rignot wrote in Science that West Antarctica's Pine Island Glacier was retreating at 1.2 kilometers a year (plus or minus 0.3 kilometers) and that its ice was thinning 3.5 (plus or minus 0.9) meters per year. "The fast recession of the Pine Island Glacier, predicted to be a possible trigger for the disintegration of the West Antarctic Ice Sheet, is attributed to enhanced basal melting of the glacier's floating tongue by warm ocean waters," Rignot wrote. This glacier is widely believed to be "the ice sheet's weak point". While the accelerated melting of this glacier does not portend an immediate disintegration of the West Antarctic Ice Sheet, Alley wrote that "most models indicate that the retreat would speed up if it kept going". One observer was quoted in this context as stating that a quick collapse of the West Antarctic Ice Sheet would "back up every sewer in New York City". Rignot speculated that warmer ocean waters were causing Pine Island's rapid bottom melting. "This is one of the most sensitive ice sheets to climatic change. For many, many years we have neglected the importance of bottom melting," Rignot said.

"The sudden appearance of thousands of small icebergs suggests that the shelves are essentially broken up in place and then flushed out by storms or currents afterward," said Scambos. The Larsen and Wilkins ice shelves have been melting since the 1950s, and scientists had expected them to fall apart, but the disintegration occurred more quickly than anticipated. "We have evidence that the shelves in this area have been in retreat for 50 years, but those losses amounted to only about 7,000 square kilometers," said David Vaughan, a researcher with the Ice and Climate Division of the British

Antarctic Survey. "To have retreats totaling 3,000 square kilometers in a single year is clearly an escalation. Within a few years, much of the Wilkins ice shelf will likely be gone".

Speculation regarding the future of the West Antarctic Ice Sheet takes place in a climatic context which includes paleoclimate records and instrument readings indicating that the western Antarctic Peninsula has warmed significantly during the last century, with the trend accelerating at the end of the period. Temperatures in this area rose four to five degrees C. during the century, according to the World Wildlife Fund. Temperatures in the area averaged above longer-term averages more than 75 percent of the time during the last third of the twentieth century. The number of annual days above freezing has increased by two to three weeks, mainly during the most recent quarter century. This temperature increase tends to erode coastal ice sheets. "Large areas of ice shatter as the melt water percolates into fractures, and deep cracks are forced open to the base of the ice sheet by the weight of the water. Once ice sheets weaken to a critical point, they may collapse very suddenly".

Average temperatures have increased by almost five degrees F. during the last 50 years on the Antarctic Peninsula, an arm of the mainland reaching a thousand miles northward toward the southern tip of South America. Glaciers are in retreat throughout the peninsula, melting ice that may be tens of thousands of years old. Woodard described the scene atop one Antarctic glacier as chunks of it calved into the sea:

Andy Young is drilling holes into the melting snow on top of the Marr Glacier when it happens. There's a sharp crack followed by a rumbling thunder as garage-size chunks of blue ice calve off a nearby glacial cliff and tumble into the frozen sea 400 feet below, throwing waves across the harbor. "That's a good one," Mr. Young says as he sticks a flag marker into the newly drilled hole. "You can see why we mark the route up the glacier you just don't want to get too near the edges."

Woodard reported that year by year more rocky surfaces emerge from beneath the melting walls of ice behind Palmer Station, a United States research facility operated by the National Science Foundation which is staffed by about 40 people. Scientists absent for a year find new beaches, outcroppings, even islands that had been hidden for thousands of years under the ice.

For several decades, scientists at the Palmer Station, astride the northwestern edge of Antarctica's Palmer Peninsula, have watched steady changes in the environment, as well as the flora and fauna it supports. Few of them doubt that many of the changes they observe are caused by a warmer climate. Annual average temperatures at the station have risen three to four degrees F. since the 1940s; winter averages have climbed seven to nine degrees during the same period. Southern elephant seals, which can weigh as much as 8,800 pounds, usually raise their young near the Falkland Islands to the north, but, in recent years, several hundred of them have been living year-round on the Palmer Peninsula. Fur seals, virtually unknown near the Palmer Station before 1950, have colonized the area around the station in the thousands, as areas once covered by ice and snow year-round have been sprouting "low grass, tiny shrubs, and mosses".

At the same time, animals that depend on sea ice for food have been migrating southward. Adelie penguins (18-inch birds which look as if they are wearing tuxedoes) have become scarcer near the Palmer Station because they eat krill which arrives with sea ice. Adelie populations within two miles of the Palmer Station have declined by about 50 percent in 25 years, including a 10 percent decline during two years in the late 1990s. About 1950, four of five winters at Palmer Station produced extensive sea ice; by the end of the century, the average was two winters in five. The Adelie penguins have been moving their range southward toward areas which heretofore had been too cold and barren for them.

Researchers say the climate near the ice shelves has warmed roughly 4.5 degrees F. since the 1940s, causing surface ice to melt. That's just the tip of the iceberg, so to speak. As this surface ice melts, it forms deep pools of water which forces its way into cracks in the ice shelf. "The weight of the water essentially forces the cracks open, so a relatively small amount of climate warming can destroy a large, centuries-old ice shelf" Scambos said. Scambos' research team is unsure whether the localized warming is part of a global warming trend, or whether it is simply a normal fluctuation in regional climate.

Scientists who study this situation have no firm estimates regarding how much warmer the earth would have to become to provoke a general collapse of the West Antarctic Ice Sheet. Some give it six degrees C. (10 degrees F.) and a century or two. Others believe the West Antarctic Ice Sheet as a whole may not drain into the sea for several centuries, if at all. Uncertainty regarding the future of the West Antarctic Ice Sheet has introduced a great deal of variability into models which estimate much global warming may cause world-wide sea levels to rise.

A study by James G. Titus and Vijay Narayanan examines some of the reasons why estimates of sea-level rise vary so widely. Titus and Narayanan express this wide range of variables in terms of probabilities:

The reviewer assumptions imply a 50 percent chance that the average global temperature will rise 2 degrees C., as well as a 5 percent chance that temperatures will rise 4.7 degrees C. by 2100. The resulting impact of climate change on sea level has a 50 percent chance of exceeding 34 centimeters and a 1 percent chance of exceeding one meter by the year 2100, as well as a 3 percent chance ofa2meter rise and a 1 percent chance of a four-meter rise by the year 2200.

Assuming that non climatic factors do not change, there is a 50 percent chance that global sea level will rise 45 cm, and a 1 percent chance of a 112 cm rise by the year 2100; the corresponding estimates for New York City are 55 and 122 cm.

The speed with which Antarctic ice may melt depends not only on how much temperatures rise, but also on the ways in which ice moves within the ice cap. Jonathan L. Bamber, David G. Vaughan, and Ian Joughin have been studying these "rivers" of subsurface Antarctic ice. "It has been suggested," they write, "that as much as 90 percent of the discharge from the Antarctic Ice Sheet is drained through a small number of ice streams and outlet glaciers fed by relatively stable and inactive catchment areas." Their research suggests that "each major drainage basin is fed by complex systems of tributaries that penetrate up to 1,000 kilometers from the grounding line to the interior of the ice sheet." Such "complex flows" are noted throughout the Antarctic ice sheet by these researchers.

Bamber, Vaughan, and Jouglin assert that "this finding has important consequences for the modeled or estimated dynamic response time of past and present ice sheets to climate forcing". The researchers also find evidence of similar (although smaller in scale) ice-sheet dynamics in Greenland. "This evidence," they write, "challenges the view that the Antarctic plateau is a slow-moving and homogenous region". These researchers also contend that the dynamics of large ice flows are too complex for present-day models to predict, so climate modelers have very little idea how global warming will affect the largest of the Earth's remaining ice masses.

An iceberg twice the size of Delaware (183 by 22 miles) broke off the Ross Ice Shelf during the third week of March 2000, moving the boundary of the Ross Ice Shelf southward by 40 miles. Two and a half weeks later, on April 8, another piece, this one roughly twice the size of Manhattan Island, calved from the Ross Ice Shelf.

While icemelt on the Antarctic Peninsula is definitely being produced by climatic warming, the case for global warming as a factor in the calving of the Ross Ice Shelf is rather tenuous. "I would lean against global warming as an explanation," Michael Oppenheimer of the Environmental Defense Fund was quoted as saying. Scambos told USA Today that periodic calving of the Ross Ice Shelf has changed the overall size of the ice mass in the area little since it was first mapped early in the twentieth century. While the breakup of ice sheets on both sides of the Antarctic Peninsula has been provoked by climatic warming, calving of the Ross Ice Shelf seems to be related to the usual dynamics of glacial formation and movement. The Ross Ice Sheet is much closer to the South Pole than the Antarctic Peninsula, with a climate that remains well below freezing for the entire year, even as climatic warming erodes ice sheets on the Antarctic Peninsula.

Global Warming Reduce Sea Levels

According to two studies, a doubling of precipitation over Greenland and Antarctica would lower mean sea level between 1.3 millimeters per year and 4.2 millimeters per year. P. Huybrechts and J. Oerlemans conclude that additional precipitation at higher temperatures will increase the mass of the Antarctic Ice Sheet and thereby lower sea levels.

Considered alone, Titus and Narayanan write, warmer temperatures in the Antarctic could increase snowfall and slightly lower worldwide sea levels. They expect Antarctic air temperatures to rise by approximately 2.5 degrees C. during the next century, largely as a result of reduced sea ice. For each degree of warming, they expect Antarctic precipitation to increase about eight percent, "equivalent to a 0.4 millimeters per year drop in sea level". They conclude,

Antarctica's temperatures are well below freezing; so unlike Greenland, warmer air temperatures will not cause a significant degree of glacial melting. Warmer water temperatures, by contrast, could potentially increase melting of the marine-based West Antarctic Ice Sheet and adjacent ice shelves.

Titus and Narayanan caution, however, that assertions linking warming to increasing snowfall fail to examine the possibility that "warmer circumpolar ocean water will intrude beneath the ice shelves, increase their rates of basal melting, decrease the backpressure that they exert on the ice streams, and thereby accelerate the rates at which ice streams convey ice from the Antarctic interior toward the oceans". They maintain that some of the recent melting along the edges of the West Antarctic Ice Sheet can be attributed to this mechanism. According to Titus and Narayanan, any warming of the circumpolar ocean is likely to lag behind the general increase in global temperatures by at least 50 years, and perhaps by a century or more.

Even with a large rate of ice-shelf melting in Antarctica, Titus and Narayanan believe, The Antarctic contribution to sea level may be negligible. Because ice shelves float and hence already displace ocean water, shelf melting would raise sea level only if it accelerates the rate at which ice streams convey ice toward the oceans. Several models suggest, however, that shelf melting will not substantially accelerate ice streams and even the models that project such an acceleration generally suggest a lag of a century or so. Thus, through the year 2100, we estimate a 60 percent chance that the sea-level drop caused by increased Antarctic precipitation will more than offset the sea-level rise caused by increased ice discharge; this probability declines to 50 percent by 2200. Our analysis suggests that if Antarctica is going to have a major impact on sea level, it will probably be after the year 2100.

Arctic icemelt

Every winter since 1912, when an iceberg claimed the Titantic in the North Atlantic, the U.S. Coast Guard has issued warnings for an average of 500 icebergs which float south of 48 degrees north latitude, calving from Greenland's ice sheet. These icebergs threaten the shipping lanes of "Iceberg Alley," the Grand Banks southeast of Newfoundland. During the El Nino winter of 1998–1999, for the first time in 85 years, the Coast Guard did not issue a single iceberg alert for that area. "The lack of ice is remarkable," said Steve Sielbeck, commander of the International Ice Patrol.

Rapid icemelt in the Arctic was evident to some observers by the 1980s. By 1989, the Arctic ice was reported to be thinning quickly. The Scott Polar Institute of the United Kingdom reported that ice in an area north of Greenland had thinned from an average of 6.7 meters in 1976 to 4.5 meters in 1987. At about the same time, a British Antarctic survey base reported unusual losses of ice as well.

Thomas V. Lowell, writing in the Proceedings of the National Academy of Sciences, cites "fieldwork on the western margin of the Greenland Ice Sheet which provides evidence that within the last 150 years, 94 percent of the lobes along the western edge of the Greenland ice sheet withdrew, whereas 70 percent of the independent glaciers retreated". Lowell, finding that glacial melting began before massive infusions of greenhouse gases into the atmosphere due to human activity, raises the possibility that twentieth-century warming may have many causes, only one of which is human-induced greenhouse forcing. Mark B. Dyurgerow and Mark F. Meier report that glaciers around the world began to melt, on balance, most recently about 1975, with "further acceleration in the last decade the 1990s".

National Aeronautics and Space Administration scientist Bill Krabill leads a team which began measuring the mass of the Greenland Ice Sheet during the early 1990s. Between 1993 and 1998, Krabill's team, flying as many as 2,000 miles a day over the ice cap, compiled the first detailed "inventory" of the Greenland Ice Sheet, from which they may be able to deduce whether it is changing and, if so, how. The team's surveys indicate, "The ice in southeastern Greenland, a region long thought to be frozen tight, has thinned rapidly. The ice in some regions has shrunken as much as thirty-three feet. The ice balance for the entire southern half of Greenland is in the red". Snowfall has increased ice mass in some inland locations of Greenland, especially its northern half, as warmer temperatures allow the import and convection of more moisture there.

The studies by Krabill and colleagues indicate that three areas of southern interior Greenland have been accumulating icepack at rates of up to 10 inches per year. In the outer regions of the ice sheets, however, the researchers reported that large areas of ice have been thinning, with the rate of thinning increasing rapidly towards the ocean. The most rapid thinning rates (one to three feet per year) were observed in the lower depths of Greenland's southeast coast outlet glaciers, the researchers reported. Ice is thinning to some degree at most elevations of Greenland below 2000 meters above sea level. At elevations higher than 2,000 meters, an average thickening of the ice sheet of 0.5 to 0.7 centimeters a year was measured from 1993 to 1998. The researchers noted that areas of thinning in the eastern part of Greenland also experienced warmer-than-normal temperatures between 1993 and 1998. "However, we also observe areas of thinning near the West coast, where many locations were cooler than normal," the researchers reported.

Krabill's work suggests that Greenland's southeastern glaciers are thinning rapidly and that their lower elevations may be particularly sensitive to potential climate changes. "The results of this study are important in that they could represent the first indication of an increase in the speed of outlet glaciers," said Krabill. An outlet glacier acts as a major ice drainage region for an ice sheet.

"The excess volume of ice transported by these glaciers has had a negligible effect on global sea level thus far, but if it accelerates or becomes more widespread, it would begin to have a detectable impact," Krabill said. Researchers reported that the glacial thinning is too large to have resulted solely from increased ice-surface melting or decreased snowfall. The researchers believe the thinning is the result of increasing discharge speeds of glaciers flowing into the Atlantic Ocean. Krabill said surface-melt water might be seeping to the bottom of glaciers. Such seepage may be reducing the friction between the ice and the rock below it, enabling the glaciers to slide with less friction across the bedrock and thus allow more ice to slip into the ocean, according to Krabill.

Krabill elaborated,

The results of this study are significant because they provide the first evidence of widespread thinning of low-elevation parts of one of the great polar ice sheets. The results also suggest that the thinning outlet glaciers must be flowing faster than necessary to remove the annual accumulation of snow within their basins.

"Why they are behaving like this is a mystery," said Krabill, "but it might indicate that the coastal margins of ice sheets are capable of responding quite rapidly to external changes". One such external stress could be global warming, Krabill said.

A new aerial survey was summarized by Krabill and colleagues in the July 21, 2000 edition of Science. "Approximately 51 cubic kilometers of ice per year [is melting] from the entire [Greenland] ice sheet, sufficient to raise global sea level by 0.005 inches per year, or approximately seven percent of the observed rise," said Krabill. "Why the ice margins are thinning so rapidly warrants additional study. It may indicate that the coastal margins of ice sheets are capable of responding more rapidly than we thought to external changes, such as a warming climate".

An Oregon State University study suggests that North American polar temperatures will rise more than eight degrees C. (14 degrees F.) with a doubling of atmospheric carbon dioxide levels. The polar regions warm more than the mid-latitudes or tropics for two reasons: Carbon dioxide produces more heat absorption in areas with relatively low amounts of water vapor, where relatively low temperatures do not allow the air to hold as much water vapor as it does in warmer regions. In addition, the melting of snow and ice alters the reflective landscape dramatically; ice and snow, which reflect most of the sunlight reaching the surface, are replaced by radiation-absorbing liquid ocean and bare land.

By the end of the twentieth century, pronounced warming was leaving its mark in parts of the arctic. Anecdotal evidence of global warming is striking in the area. For example, scientists keeping records at Toolik Lake, north of Alaska's Brooks Range, noted that the lake's surface temperature rose by almost three degrees C. between 1979 and 1994. A team of NOAA researchers found an increase in surface temperatures of 5.5 degrees C. (about nine degrees F.) during 30 years, between 1965 and 1995, at nine stations north of the Arctic

Circle. One report indicates that average air temperatures directly above the snowy top of the Greenland Ice Sheet rose as much as 18 degrees F. during the 1990s, with the year 2000 the warmest year. "There has been no trend in the summer maximum temperature," said Konrad Steffen of the University of Colorado, "but enormous fluctuations and warming trends in the fall, winter and spring".

Evidence of rapid warming had become part of everyday life in the Arctic by the late 1990s. Point Barrow, Alaska, broke its annual average temperature record in 1998 by three degrees F., a very large number. The old annual record, set in 1940, had been 14 degrees F. Point Barrow's annual average temperature in 1998 was seven degrees F. higher than the 30-year average (1961–1990), which meteorologists use as a benchmark. One day during the summer of 1998, the high temperature at Point Barrow hit a balmy 67. Many of the 4,000 people who live in Point Barrow, most of them Inuit, express fears that lack of ice along the shore may expose their storm-lashed town to flooding. They're also afraid that the bowhead whales, which provide the core of their diet, may swim further offshore, making them harder to kill. Whale meat stored in holes outdoors also may be ruined by an untimely thaw. During the summer of 1998, some melting occurred as liquid water seeped into some storage areas.

Average temperatures at the mouth of the Mackenzie River were nine degrees F. above long-term averages during 1998. Several buildings have been lost to erosion by seawater in the village of Tuktoyaktuk, near the mouth of the Mackenzie. Several decades ago, miners in the area deposited toxic wastes in ponds which were expected to remain frozen (and the toxic materials sealed by the permafrost). With warmer temperatures, some of these toxic dumps may thaw and leak.

A significant rise in Siberian temperatures during the late twentieth century has been provoked by "a speeding up of atmospheric circulation in the Northern Hemisphere which results in more warm air being transported deep into the region". Tundra and taiga (high-latitude forests in Asia) have been moving northward during the late twentieth century, as satellite surveys indicate that the amount of photosynthesis between 45 and 70 degrees north latitude has increased. This trend is linked by William J. Burroughs to a "rise in spring temperatures, which lengthened the growing season at high altitudes".

In this context, Burroughs raises one of many perplexing questions regarding how the Earth manages carbon:

The possible increase in the growth of the taiga touches on another interesting aspect of the current global warming, which relates to our limited knowledge. Measurements in the early 1990s suggested that the amount of carbon dioxide being absorbed at high altitudes was much greater than previously estimated. This "Great Northern Sink" has been the subject of a great deal of speculation among scientists.

This point of view is not universally accepted among scientists. Witness a nearly opposite assertion by tundra researcher George W. Kling, an associate professor of biology of the University of Michigan, who said, "Our latest data show that the Arctic is no longer a strong [net] sink [absorber of] carbon. In some years, the tundra is adding as much or more carbon to the atmosphere than it removes".

The thinning of ice in the Arctic affects life all along the food chain. When the ice breaks up earlier than usual, polar bears have less time to build fat reserves, and must rely on these reduced reserves for a longer period of time before ice forms again in the fall. Calculations indicate that a mean air temperature increase of one degree C. could advance the date of ice breakup nearly a week in western Hudson Bay. Hunger among adult bears will be reflected in their offspring, who will be born smaller and more prone to die at an early age.

In the Bering Sea and in Hudson Bay, evidence of stress in polar bear populations is mounting as sea ice retreats. Some native people in the Arctic have reported difficulty in hunting marine mammals such as walrus in recent years as sea ice has diminished.

On October 2, 1997, the Canadian Coast Guard icebreaker Des Groseilliers entered the pack ice of the Arctic Ocean about 250 miles due north of Prudhoe Bay, Alaska. Starting at this position, the Des Groseilliers began a year-long voyage, drifting with the pack ice. Ice Station SHEBA traveled about 1,000 miles in 13 months, ending its journey about 450 miles northwest of Barrow, Alaska. The icebreaker was a base of operations and home for a group of more than 50 scientists involved in the scientific research program called SHEBA,

Surface Heat Balance of the Arctic Ocean. Scientists on the ship conducted tests to determine the physics of ice melt, and were astounded by what they found. The ice was much thinner, as a rule, than their models had led them to believe. Where they had expected to find ice six to eight feet thick, it was three to four feet thick. The ocean underneath the ice also was warmer and less saline than it had been previously.

The lack of salinity in the upper reaches of the ocean under the ice cap indicated that a great deal of ice had melted in a relatively short time. The speed with which the Arctic ice cap is thinning has led some observers to estimate that the entire body of ice (an area the size of the continental United States) will melt during the summer by roughly the year 2020. "If the current trend continues, the Arctic ice pack will disappear," Michael Ledbetter of Ice Station SHEBA said. "The models say it and we didn't find anything at SHEBA that would cause us to modify those predictions. In twenty-five years, we might lose the entire Arctic Ocean ice pack".

During late 1999, a team of scientists reported in Science that the Arctic ice cap had lost a Texas-sized measure of its mass during the previous two decades. The team, led by Konstantin Vinnikov of the University of Maryland, had to inventory many retreats in the ice cap across the Arctic Ocean. The trend the scientists found was unmistakable, with its human origins nearly as clear. The team set the chance that such melting is a measure of natural climate variability at two percent. Their study "strongly suggests that the observed decrease in Northern Hemisphere sea ice is related to human caused global warming".

During the second week of August 2000, a patch of open water a mile wide was reported at the North Pole. Initial reports indicated that this might have been the first melting of ice at the pole since the Eocene Period, about 50 million years ago. "It was totally unexpected," said Dr. James J. McCarthy, an oceanographer, director of the Museum of Comparative Zoology at Harvard University and the coleader of a group working for the Intergovernmental Panel on Climate Change.

McCarthy took part in a cruise of the Arctic aboard the Russian icebreaker Yamal. On a similar cruise six years ago, he recalled, the icebreaker plowed through an icecap six to nine feet thick at the

North Pole. The Yamal cruised six miles away to find ice thick enough for the 100 passengers to get out and be able to say they had stood on the North Pole, or close to it. They saw ivory gulls flying overhead, the first time ornithologists said they had been sighted at the pole.

The iceless North Pole was widely publicized as a new indicator of global warming, drawing a counterattack from climate-change skeptics. S. Fred Singer said in the Wall Street Journal that he had cruised the Arctic with the U.S. Navy, and "I can testify that icebreakers always search for leads [cracks] to make their way through the ice" (Singer 2000, A-18). Peter Wadhams, director of the Scott Polar Institute in Cambridge, England, said, "Claims that the North Pole is now ice-free for the first time in 50 million years [are] complete rubbish, absolute nonsense. What is happening is of concern but it is gradual, not sudden or stupendous". Wadhams said that the extent of Arctic open water during the summer months is on the rise, as Arctic ice gradually thins. "In an ice pack, there will always be some open water" this time of year even at the North Pole, said Donald Perovich, who studies links between polar ice and climate at the U.S. Army Corps of Engineers' Cold Regions Research and Engineering Laboratory in Hanover, New Hampshire. The trend in the Arctic is rather dramatic over several decades, however. The polar ice cap lost about 40 percent of its volume between 1958 and 1997, according to figures assembled by researcher Drew Rothrock and colleagues. They compared ice soundings taken by U.S. Navy submarines between 1958 and 1976 with soundings taken during scientific submarine cruises from 1993 to 1997.

The same month that the passengers of the Yamal found open water at the North Pole, an aluminum hulled catamaran owned by the Royal Canadian Mounted Police, the St. Roch II, crossed the 900-mile channel through Canada's Arctic islands in 30 days, a record. The crossing of the Roch II, made with minimal delay by ice, raised speculation that regular summer shipping through the Arctic would soon cut 5,000 miles off the shortest route between Europe and Asia, and open the once-fabled Northwest Passage sought by mariners for 500 years.

Water temperatures in parts of the Arctic Ocean increased an astonishing 1.6 degrees C. during the decade of the 1990s alone. If the Arctic continues to warm at anywhere near this rate, by the year 2020 the Northwest Passage may open through the thawing Arctic Ocean year-round for the first time in human history. The passage was sought in vain by early sixteenth century explorers of North America. Construction of the Panama Canal finally allowed shorter oceanic transport of goods between the coasts of North America. Canadian military, police, and customs officials already have begun to plan ways to use the passage, as they examine how to manage the new waterway.

Woodard described the retreat of ice sheets in Iceland during the last century:

From the gravelly, newborn shores of this frigid lagoon, Iceland's Vatnajokull ice cap is breathtaking. The vast dome of snow and ice descends from angry clouds to smother jagged 3,000-foot-tall mountains. Then it spills out from the peaks in a steep outlet glacier 9.3 miles wide and 12.4 miles long an insignificant appendage of Europe's largest ice cap despite its impressive size. A hundred years ago, however, there was no lagoon here. The shoreline was under one hundred feet of glacial ice. The outlet glacier, known as Breidamerkurjokull, extended to within 250 yards of the ocean, having crushed medieval farms and fields in its path during the preceding centuries, a time now referred to as the Little Ice Age. Today, Breidamerkurjokull's massive snout ends about two miles from the ocean. In its hasty retreat, the glacier has left the rapidly expanding lagoon, which is filled with icebergs calved from its front. The lagoon is 350 feet deep and has nearly doubled its size during the past decade. Every year, it grows larger, threatening to wash out Iceland's main highway.

Speed of Polar Climate Change

Evidence from ice cores taken in Greenland has altered the scientific view regarding the speed with which climatic change takes place, especially at higher latitudes. The older assumption that climate changes very slowly in the polar regions is built into the English language when we say that something moves with "glacial speed." On the contrary, the ice-core record indicates that several rapid warming and cooling have convulsed the Arctic during the last 100,000 years.

The last ice age may have ended much more quickly than previously thought. Jeffery P. Severinghaus of the Scripps Institute of Oceanography described a new method of analyzing gases trapped in Greenland's ice sheet which indicates that temperatures in the area rose about 16 degrees F. within a decade or two at the end of the last great ice age. "The old idea was that the temperature would change over a thousand years, but we found it was much faster," said Severinghaus. "We know that over the next one hundred years the Earth will probably warm because of the greenhouse effect," Severinghaus continued. "There is a remote possibility that we might trigger one of these abrupt climate changes. This certainly gives us pause".

Rapid climate changes within a few thousand years are sketched by Jonathan Adams, Mark Masli, and Ellen Thomas:

Ice-core evidence indicated that the Eemian was punctuated by many short-lived cold events, as shown by variations in electrical conductivity (a proxy for windblown dust, with more dust indicating colder, more arid conditions) and stable oxygen isotopes (a proxy for air temperature) of the ice that were used by these workers to infer climatic conditions. The cold events seemed to last a few thousand years, and the magnitude of cooling was similar to the difference between glacial and interglacial conditions; a very dramatic contrast in climate. Furthermore, the shifts between these warm and cold periods seemed to be extremely rapid, possibly occurring over a few decades or less. Sudden and short-lived warm events occurred many times during the generally colder conditions that prevailed between 110,000 and 10,000 years ago.

The last cold "spike" (or "Heinrich Event"), the Younger Dryas, peaked between 13,000 and 11,500 years ago. The sudden beginning and end of the Younger Dryas has been studied in particular detail utilizing ice-core and sediment records. A detailed study of two Greenland ice cores suggests that the main Younger Dryas-to-Holocene warming took several decades in the Arctic, but was marked by a series of warming surges, each taking less than five years. About half of the warming was concentrated in a single period of less than 15 years. A rapid global rise in atmospheric methane concentration that occurred at the same time suggests that the warmer and more humid climate (causing more methane output from swamps and other biotic sources) was a factor in these rapid changes. Scientists are at a loss to explain why the climate can change suddenly, as they describe records that indicate such changes:

It is still unclear how the climate on a regional or even global scale can change as rapidly as present evidence suggests. It appears that the climate system is more delicately balanced than had previously been thought, linked by a cascade of powerful mechanisms that can amplify a small initial change into a much larger shift in temperature and aridity.

Past rapid changes in regional climates took place without human provocation. Factors contributing to these rapid changes may have included variations in the sun's radiance, changes in angle of the Earth's axis, natural variations in atmospheric levels of carbon dioxide and methane (among other "trace" gases), changes in atmospheric water-vapor levels, and dust ejected from volcanoes and other sources. Jonathan Adams and associates comment,

Climate has a tendency to remain quite stable for most of the time and then suddenly "flip," at least sometimes over just a few decades, due to the influence of various triggering and feedback mechanisms. Such observations suggest that even without anthropogenic climate modification there is always an axe hanging over our head, in the form of random very large-scale changes in the natural climate system; a possibility that policy makers should perhaps bear in mind with contingency plans and international treaties designed to cope with sudden famines on a greater scale than any experienced in written history. By starting to disturb the system, humans may simply be increasing the likelihood of sudden events which could always occur. To paraphrase W.S. Broecker; "Climate is an ill-tempered beast, and we are poking it with sticks."

Cooler Europe in a Warmer World

The Gulf Stream emerges in the ocean-atmosphere dynamics of the northern Atlantic Ocean as a crucial regulator of climate, especially for western Europe. Sediment records have produced evidence that a massive outflow of debris north to south across the North Atlantic produced by rapid melting of the Arctic ice cap (including glaciers in Greenland and northern Canada) could radically alter the climate of western Europe, producing colder winters in an otherwise warmer world.

Through most of its recorded history, Europe has experienced an unusual degree of warmth for such a northerly latitude because the Gulf Stream flows near the continent's western shores, carrying relatively warm water from tropical regions near North America. Land-locked areas at 50 degrees north latitude, similar to that of London, include the Canadian prairies and much of Siberia.

Climate models suggest that a rapid outflow of cold, relatively fresh water might divert the Gulf Stream southward, depriving Europe of oceanic warmth that bathes the continent during most winters. During such periods in the past, Europe may have endured a harsher climate, with less precipitation and much colder winters. Summers may have been as warm as or warmer than today, because the Gulf Stream tends to dampen some of continental warming 's effects during the summer months.

The northern Atlantic Ocean is a region of particular importance to the global climate system because it contains one of a small number of sites in the world where down welling creates deep-ocean water. At this site, northeast of Iceland, surface seawater is cooled until, because of its increased density, it sinks and forms a narrow, deep current the "North Atlantic Deep Water" which hugs the eastern shore of Greenland.

While a geological-sciences doctoral student at the University of Colorado's Institute of Arctic and Alpine Research in Boulder, Don C. Barber made a case that a sudden surge of cold weather was triggered in Greenland and Europe 8,200 years ago by a massive flow of cold, fresh water from the Hudson Bay region into the North Atlantic Ocean. According to Barber, Hudson Bay (now salt water) housed the world's largest freshwater lake system, which formed from melting ice following the last ice age. An ice dam which had been holding the fresh water broke, spilling much of it into the North Atlantic Ocean, suppressing the Gulf Stream southward. Barber said that Greenland ice cores show that temperatures dropped by as much as 15 degrees F. in central Greenland and by nearly six degrees F. in western Europe following the lake drainage. Barber and colleagues estimated that when the ice dam from a remnant of the Laurentide Ice Sheet collapsed, the lake water flowing through the Hudson Strait into the Labrador Sea flowed out at a rate about 15 times greater than the present flow of the Amazon River.

According to the Times of London, evidence that even a small change in ocean currents may sharply decrease temperatures in Europe has been found in seabed sediment by Nick McCave, a professor of geology at Cambridge University. These deposits, collected south of Iceland, show that changes in currents associated with the Gulf Stream have coincided with large shifts in climate, including the "Little Ice Age," which plagued Europe roughly between 1400 and 1800, a time when persistent cold weather caused many harvests to fail and Londoners held fairs on the frozen Thames.

William J. Burroughs describes the dynamics of open-ocean circulation which could produce a significant cooling in Europe as the rest of the world warms:

Modeling work suggests that circulation patterns are extremely sensitive to the runoff of rainwater from the continents, the number of icebergs calved off Greenland, and the amount of precipitation from low-pressure systems tracking northeastwards past Iceland, and into the Norwegian Sea. Small variations in the total output may be able to trigger sudden switches in global ocean currents.

"Drunken Forest"

Permafrost is profoundly destablilized by warming above the freezing point of water. Permafrost regions which are threatened by global warming also are being provoked to premature thawing by other human activities, including mining, forestry, and the construction of pipelines to carry oil and other fuels. "When vegetation is removed, the permafrost is far more likely to melt and cause flooding," Burroughs writes.

Warming and eventual thawing of permafrost will have major environmental and societal impacts. As tundra and permafrost thaw, they release stored carbon dioxide and methane which raises the atmosphere's level of greenhouse gases. E.A. Koster and M.E. Nieuwenhuyzen comment in Greenhouse-Impact on Cold-Climate Ecosystems and Landscapes,

Where permafrost is ice-rich, or contains massive ground ice, subsidence and settling due to thawing will eventually occur. This could amount to several meters, with severe consequences for roads, buildings, bridges, pipelines, and other structures. This can alter drainage patterns, and change the course of streams. Some areas may become swamp-like.

The conditions Koster describes already have been observed in some Alaskan forests. Large patches of forest were described in newspaper reports of the late 1990s as drowning and turning gray as the thawing ground sank under them. Trees and roadside utility poles, destabilized by thawing, lean at crazy angles. The warming has contributed a new phrase to the English language in Alaska "the drunken forest." Scientists at the University of Alaska say that temperatures in the same region have risen about five degrees F. during the past 30 years.

Human settlements may be damaged as permafrost melts around the Bering Sea. Such thawing could cause the land to subside as much as five meters, or more than 15 feet. The thawing of permafrost is also expected to destroy caribou habitat, cause landslides and erosion, clog salmon spawning rivers with silt, and trigger the loss of some boreal forests.

Between 20 and 25 percent of the land in the Northern Hemisphere experiences at least seasonal permafrost. Of this area, Koster and Nieuwenhuyzen write, "At the moment, Arctic tundras provide a major sink of CO2. Changes in vegetation, as well as in the organic matter within the active layer, induced by climatic change, may cause this sink to become a major source, thus accelerating climatic changes" A large amount of carbon is stored in peat deposits, most of which have accumulated since the last glaciation. "A climatic shift to higher temperatures will increase the release of CH4 [methane] from deep peat deposits, especially from tundra soils. Likewise, the release of CO2 will increase, but hardly by more than 25 percent of its present level".

Nils Malmer, a plant ecologist in the Department of Ecology, Lun University, Sweden, describes how warmer temperatures gradually convert peatlands from a carbon sink to a carbon source:

Virgin peatlands have generally been suggested to have a net accumulation of organic matter and are therefore treated as carbon sinks in most studies of the global carbon cycle. When peatlands are drained for agriculture or forestry, a net release of carbon takes place through an increasing decomposition of organic matter. For a short period of time (less than 100 years) this release can be at least partially compensated by increased tree biomass on drained sites. The instantaneous release of carbon from a peatland is, of course, highly increased when a peat deposit is exploited and used as fuel.

Slip-Sliding Glaciers

Ice-core samples taken from high elevations in the Himalayas indicate that the 1990s were the warmest decade of the last 1,000 years at "the roof of the world." A team of scientists drilled holes at roughly 500-foot elevation intervals in the ice on a flank of Xixabangma, which rises to 26,293 feet. "This is the highest climate record ever retrieved, and it clearly shows a serious warming during the late twentieth century, one that was caused, at least in part, by human activity," said Lonnie Thompson, a professor of geological sciences at Ohio State University, who led the study. The ice cores also contain evidence of major droughts, including one that began in 1790, which caused the deaths of more than 600,000 people in India, where monsoon rains did not arrive for six years. The research team also found evidence that levels of atmospheric dust have quadrupled over the Himalayas during the twentieth century. Concentrations of chlorine have doubled during the same period.

Global warming will raise snow levels and generally will cause seasonal snow packs to melt more quickly. M. Zimmermann and W. Haeberli, who have studied mountain debris flow in their native Switzerland, assert, Glaciers usually protect subglacial sediments from instability due to increased normal pressure and sheet strength. Glacier retreat has caused remarkable changes in the cryosphere of the Alps, which resulted in an extension of zones prone to debris-flow initiation. The melting of small glaciers in particular exposes steep moraines. In addition, many small lakes have formed behind retreating glacier margins. Breaking of such natural dams can cause disastrous debris flows. Continued glacier shrinkage would expose even more morainic material and, hence, further expand the area of potential debris-flow initiation. The reduction of summer runoff due to decreasing glacier surface is not likely to compensate for this negative effect.

Switzerland's Grindewald Glacier has retreated about 1.5 kilometers since 1850, with about half the retreat since 1940. The mean surface temperature of alpine permafrost in Switzerland rose between 1.0 and 1.5 degrees C. from 1880 to 1950, then continued more-or-less stable from 1950 to 1980, rising again after that. Warming mountain temperatures have caused "upward displacement" of plant species on 26 alpine peaks, according to one study and 30, according to another.

As glacial ice melted in the higher elevations of the Alps during the summer of 2000, long-lost debris and human remains surfaced in unusually large amounts. On 15,800-foot Mont Blanc, the highest elevation in the Alps, fragments from an Indian Airlines Boeing 707, including human bones, surfaced at the foot of the Bossons glacier, near the ski resort of Chamonix. The airliner crashed in 1966. The remains of a Dijon woman who disappeared in October 1977 were found at 9,000 feet on the Peclet-Polset glacier during August. According to an account in the Times of London, "Bones, climbing equipment and parts from the Boeing, and another Indian airliner that crashed in 1950, had been appearing for years on the glaciers, but the heat wave had accelerated the process".

The Intergovernmental Panel on Climate Change declared in 1995 that between a third and one-half of existing mountain glacier mass could disappear in 100 years. Ice cores drilled in the Tibetan plateau indicate that the last half-century probably was the warmest period in the last 12,000 years in that area.

Ice-core records indicate that higher elevations in the tropics are warmer than they have been in at least 2,000 to 3,000 years, causing glaciers on the highest peaks to retreat. The edge of the Qori Kalis glacier in the Peruvian Andes retreated at a rate of 13 feet a year from 1963 to 1978, but by 1995, the rate of retreat was 99 feet a year. In the meantime, sea levels have risen four to ten inches during the last century, more than they had changed during the previous thousand years. The sea level is now at its highest level in 5,000 years, and rising. In Peru, the retreat of glaciers in the Andes was raising concern during the late 1990s for irrigation water which usually flows from the mountains to the coastal desert. The same rivers also are dammed for hydroelectric power.

Indian researchers have warned that glaciers in the Himalayas are melting at an alarming rate and could cause a catastrophe if melt water lakes overflow into surrounding valleys. "All the glaciers in the middle Himalayas are retreating," Syed Hasnain, of Jawaharlal Nehru University in Delhi, told New Scientist magazine. The Gangorti glacier at the head of the Ganges River is receding 100 feet a year, according to Hasnain 's four-year study. If current trends continue, all the glaciers in the central and eastern Himalayas could disappear by 2035, Hasnain said.

Glaciologists Lonnie G. Thompson and Ellen Mosley-Thompson of Ohio State University have measured the retreat of the largest glaciers in the Peruvian Andes. They calculated a rate of retreat for 1963 to 1983, and found that the rate had tripled between 1983 and 1991. The loss in volume of ice increased seven-fold in less than half a human lifetime. Between 1963 and 1987, the glacial mass on Mount Kenya in tropical east Africa decreased 40 percent. The Thompsons comment, "The loss of these valuable hydrological stores may result in major economic and social disruptions in those areas dependent upon the glaciers for hydrologic power and fresh water".

Greenhouse Forcing Break the Glacial Cycle

Anthony D. Socci has surveyed epochs of natural cooling and warming in the earth's atmosphere, with reference to levels of carbon dioxide in the atmosphere roughly 700 million years into the past, during which he finds that the earth has been both markedly warmer and colder than today. The range of temperature has remained mainly 15 degrees C. warmer or colder than today's levels. According to Socci, "The duration of anthropogenic human-caused, greenhouse warming may likely exceed 300,000 years, conceivably pre-empting at least three global cooling events normally characterized by thickening and spread of continental glaciers".

Socci believes that human-induced warming of the Earth's atmosphere is short-circuiting the glacial cycles which have dominated glacial climate for the last several-score million years. "In other words," Socci comments, "the anthropogenic contribution of carbon dioxide to the atmosphere is in excess of the volume of carbon dioxide which nature calls upon to go from one extreme of climate to the other". In raising the atmospheric level of carbon dioxide to about 370 parts per million from the pre industrial level of 280 p.p.m. in one century, humankind has broken an ages-old natural cycle.

During recent decades, study of ice cores has fundamentally changed many of science's assumptions about paleoclimate, the climate of the past. Analyzing air bubbles in ice cores of various ages, scientists can "read" a number of things, including the levels of greenhouse trace gases (notably carbon dioxide and methane) during past climatic cycles. Early ice-core work established a relationship between fluctuations in greenhouse-gas levels and temperatures. Because of ice-core records, paleoclimatologists now know that climate can change abruptly as the cycle shifts.

By 1999, a team of scientists led by J.R. Petit, drilling ice cores at Vostok, Antarctica, reached a depth of 3,623 meters, and in so doing was able to describe the composition of the Earth's atmosphere during the last four glacial cycles, or about 420,000 years. They found that the carbon dioxide level varied in a range of 180 parts per million (p.p.m.) to 300 parts per million, in a cycle which roughly parallels rises and falls in temperature. The atmospheric level of methane during the same period varied between 320 and 770 parts per billion (p.p.b.), in a similar pattern low during glacial maxima, and high during interglacial warm periods. By the year 2000, greenhouse-gas levels (about 370 p.p.m. for carbon dioxide, and 1,700 p.p.b. for methane) were much higher than any level reached for as long as humankind has proxy records from ice cores, now 420,000 years before the present time.

The Vostok ice-core investigators have not declared that human-induced warming is short-circuiting the glacial cycle. Their words are more circumspect:

A striking feature of the Vostok record is that the Holocene, which has already lasted 11,000 years is, by far, the longest stable warm period recorded in Antarctica during the last 420,000 years CO2 and CH4 concentrations are strongly correlated with Antarctic temperatures; this is because, overall, our results support the idea that greenhouse gases have contributed significantly to the glacial-interglacial change. This correlation, together with the uniquely elevated concentrations of these gases today, is of relevance with respect to the continuing debate on the future of the Earth's climate.

3

Warming Seas

Introduction

When considering the implications of global warming, usually is fixed on the third of the planet that comprises dry land. While the land warms, the other two-thirds of the Earth will be warming as well, with profound implications for the species which inhabit it, including a holocaust for coral reefs which already has begun.

According to Peter G. Brewer and associates, writing in Science, the chemistry of ocean water already is being changed by human-induced greenhouse gases in the atmosphere: "The invading wave of atmospheric CO2 has already altered the chemistry of surface seawater worldwide, and over much of the ocean, this tracer field has now permeated to a depth of more than one kilometer". Given that the ocean is slowly warming, as its carbon dioxide level rises, its capacity as a carbon sink is probably being compromised.

Ken Caldeira and Philip B. Duffy assert in Science that "uptake," or removal, of human-induced carbon dioxide by the oceans is less than many earlier investigators have assumed and that, as temperatures warm, carbon uptake will diminish further. Additionally, Caldeira and Duffy contend that absorption of carbon into the oceans of the Southern Hemisphere (the focus of their study) has been diminishing since 1880 when fossil-fuel effluvia became a factor in the composition of the atmosphere. "Ventilation of the deep

Southern Ocean was much more vigorous in the period from about 1350 to 1880 than in the recent past".

Among the important results of warming seas will be coastal erosion, shoreline inundation because of higher tide levels, higher storm surges, and saltwater intrusion into coastal estuaries and groundwater supplies. A report by the World Wildlife Fund said: "Scientific evidence strongly suggests that global climate change already is affecting a broad spectrum of marine species and ecosystems, from tropical coral reefs to polar ice-edge communities".

The level of the world's seas and oceans has been rising slowly for much of the twentieth century, a millimeter or two a year, enough to produce noticeable erosion on 70 percent of the world's sandy beaches, including 90 percent of sandy beaches in the United States. In some areas, such as the United States Gulf Coast between New Orleans and Houston, the withdrawal of underground water and oil is causing some areas to sink as the oceans rise.

More than 20 percent of the world's population lives within 30 kilometers of coastal areas. That population is increasing twice as quickly as aggregate population. According to one estimate, a one-meter rise in sea level could displace 300 million people around the world. Thirty of the world's largest cities lie near coasts, and are vulnerable to a one-meter rise in ocean level. Most of Australia's population lives in coastal cities. By 1990, more than 100 million people lived within 50 miles of coastlines in the United States.

A large number of the people who could be displaced by modest sea-level rise are residents of broad river deltas, such as the Mississippi near New Orleans, the Nile in Egypt, and the Ganges, which joins with several other rivers in Bangladesh. Half of Bangladesh itself is less than 5 meters above sea level. A one-meter sea-level rise in Bangladesh would displace about 10 to 20 percent of the human population, and cost at least six percent of Gross National Product (GNP). An ocean-level rise of three meters would affect 27 percent of the people there, and cost roughly 15 percent of GNP.

One observer of this situation believes, "Global change within the next century will be particularly hard-felt in densely populated delta areas like Bangladesh, where a substantial area is barely above sea level and vulnerable to frequent natural calamities like droughts, tropical cyclones, and tornadoes". Fesih Mahtab projects that Bangladesh's population will rise from 115 million in 1990 to 145 million in the year 2000 and 232 million in 2030. A sea-level rise of one meter would put 16 percent of the country under water.

During the twenty-first century, the phrase "environmental refugee" may become more familiar around the world. Lowland residents who could be forced out of their homes by a three-foot rise in sea levels during the next century include 26 million people in Bangladesh, 70 to 100 million in China, 20 million in India, and 12 million in the Nile Delta of Egypt. In Egypt, a one-meter sea-level rise could cost 15 percent of the country's GNP, including much of its agricultural base. A 14-inch rise in sea levels could flood 40 percent of the mudflats which ring Puget Sound, obliterating a significant habitat for shellfish and waterfowl.

The locations mentioned above are only a few of a great many examples, because the Earth's junctures of seas and rivers have been important crossroads for human trade (as well as fertile farming areas) throughout human history. Many river deltas are densely populated and very vulnerable to even a small amount of sea level rise. Forecasts of sea level rise must contend with major uncertainties. Most sea level rise projections do not indicate how much the increase hinges on what happens in Antarctica or how sudden some rises in sea level could be. William R. Cline calls some of these estimates "innocuous," as he comments, Consider sea level rise, recently downgraded from one meter to about 40 centimeters for CO2 doubling. The IPCC [Intergovernmental Panel on Climate Change] calculations are premised on no contribution to sea level rise from the Antarctic, on grounds that its temperature remains in a range (below minus 12 degrees C.) where warming increases rather than decreases glacier mass. (One reason is that warmer air is more moist, so it provides more snow for glacier buildup.) But warming of 10 degrees C. or more would cause the Antarctic to move into the domain where it contributes to sea level rise. If so, the increase in sea level would be much higher particularly if the West Antarctic ice sheet [were to begin] to disintegrate.

Rising Ocean Temperatures

In November 1994, researchers announced that they had found a 0.9 degree F. (0.5 degree C.) rise in the temperature of the deep waters of the Indian Ocean, compared to measurements taken 20 years earlier. This was the third ocean observed to be warming. In 1992, Nathan Bindoff reported that the subsurface temperature of the southwestern Pacific Ocean had increased at about the same rate, 0.9 degree F. (0.5 degree C.), over 20 years. In early 1994, a group led by Gregorio Parrilla of the Spanish Oceanographic Institute reported that the North Atlantic also was warming. In December 1994, Ed Carmack of the Institute of Ocean Sciences in Sidney, British Columbia, Canada, reported that the waters 200 to 1000 meters (660 to 3280 feet) beneath the Arctic Ocean had warmed 1.8 degrees F. (1.0 degree C.) in a decade.

Scientists at the National Oceanic and Atmospheric Administration (NOAA) reported during late March 2000 that the oceans had warmed significantly worldwide during the preceding four decades. A broad study of temperature data from the oceans showed that average water temperatures had increased from one-tenth to one-half degree, depending on depth, since the 1950s. The NOAA study uncovered significant warming between 1955 and 1995 as deep as 10,000 feet in the oceans. The greatest warming occurred from the surface to a depth of about 900 feet, where the average heat content increased by 0.56 degree F. Water as far down as 10,000 feet (in the North Atlantic) was found to have gained an average of 0.11 degrees F.

"We've known the oceans could absorb heat, transport it to subsurface depths and isolate it from the atmosphere. Now we see evidence that this is happening," said Sydney Levitus, chief of NOAA 's Ocean Climate Laboratory and principal author of the study. "Our results support climate modeling predictions that show increasing atmospheric greenhouse gases will have a relatively large warming influence on the Earth's atmosphere," said Levitus. According to Levitus, "The whole-Earth system has gone into a relatively warm state".

The ocean-temperature record compiled by Levitus and associates confirms similar trends in land-surface temperature records, with one notable period of warming between 1920 and 1940, followed by a slight cooling. By the late 1970s, both ocean- and land-surface temperatures began to spike upward to new peaks in the observational record. The magnitude of the oceanic warming surprised some experts. Dr. Peter Rhines, an oceanographer and atmospheric scientist at the University of Washington in Seattle, said it appeared roughly equivalent to the amount of heat stored by the oceans as a result of seasonal heating in a typical year. "That makes it a big number," he said.

The Environmental Protection Agency (EPA) has stated that warming seas also may be influenced by changes in the Earth's gravitational field provoked by melting ice at the poles. "Removal of water from the world's ice sheets would move the earth's center of gravity away from Greenland and Antarctica; the oceans' water would thus be redistributed toward the new center of gravity. Along the coast of the United States, this effect would generally increase sea level rise by less than 10 percent; sea level could actually drop, however, at Cape Horn and along the coast of Iceland". Climate change could also influence local sea level by changing winds, atmospheric pressure, and ocean currents, but no one has estimated these impacts.

Sea-Level Rise and Small Island Nations

On many small islands around the world, rising seas are everyday news, and global warming is on everyone's lips. Nicholas Kristof of the New York Times described how Teunaia Abeta, a resident of Kiribati island, in the Pacific, watched in horror, as a high tide came rolling in from the turquoise lagoon and did not stop. There was no typhoon, no rain, no wind, just an eerie rising tide that lapped higher and higher, swallowing up Abeta's thatched-roof home and scores of others in this Pacific Island nation. "This had never happened before," said Abeta, 73, who wore only his colorful lava-lava, a skirt-like garment, as he sat on the raised platform of his home fingering a home-rolled cigarette. "It was never like this when I was a boy."

The Alliance of Small Island States (AOSIS), including the Philippines, Jamaica, the Marshall Islands, the Bahamas, Samoa, and others, has been one of few voices at climate talks favoring swift, worldwide reductions in greenhouse-gas emissions. Their self-interest is evident: with warming already "forced," but not yet fully worked into the world's temperature equilibrium, many small island states will lose substantial territory and economic base within the next few decades.

For every other country on Earth, competing economic interests drive climate negotiations. For the small island nations, the driving force is survival. "For us, it's a matter of death and life, whereas in terms of the citizens of the industrialized countries, [global warming] will affect their lifestyles basically, but not to the extent they will be disappearing," said Bikenibeu Paeniu, Tuvalu's prime minister.

In no other place is global warming more urgent an issue, in a practical sense, than in equatorial Kiribati (pronounced "Kirabas"), a chain of coral atolls in the Pacific, near the Marshall Islands. Kiribati is a republic, a member of the United Nations, and home to about 79,000 people. Its land area, at any point, is no more than two meters above sea level. Additionally, the living corals that have raised the islands above sea level are threatened with demise by warmer ocean waters. Ierimea Tabai, who became president of Kiribati after its independence from Britain in 1979, has said, "If the greenhouse effect raises sea levels by one meter, it will eventually do away with Kiribati. In fifty or sixty years, my country will not be here".

Kiribati's 33 islands, comprising 277 square miles, are scattered across 5.2 million square kilometers (2 million square miles) of ocean, including three groups of islands: 17 Gilbert Islands, eight Line Islands, and eight Phoenix Islands. Kiribati includes Kiritimati (formerly Christmas Island), the world's largest coral atoll (150 square miles). The island of Kiribati, located roughly halfway between California and Australia, more than doubles its surface area at low tide.

During 1997, the area was devastated by El Nino, which brought heavy rainfall, a half-meter rise in sea level, and extensive flooding. About 40 percent of the atolls' coral was killed by overheated water, and nearly all of Kiritimati Island's roughly 14 million birds died or deserted the island.

Sea level rise provoked by global warming could imperil more than 3,000 small isolated islands, grouped into 24 political entities in the Pacific Ocean, with a population of about 5 million people in 800 distinct cultures. Many coral atolls also contain permanent areas of fresh water (lagoons), which are vulnerable to salinization as sea levels rise.

Central lagoons of the small islands are social centers and a source of food. According to Kristof, Children play in the water from infancy. Many adults fish or sail for a living. The Kiribati men in their loincloths go out in outrigger canoes each day to catch tuna, and the women wade out on the coral reef to dive for shellfish and net smaller fish for dinner. "People here think of the ocean as their source of livelihood, as their friend," said Ross Terubea, a Kiribati radio reporter. "It's hard to think that it would destroy us."

By 1998, rising sea levels already were swallowing some small Pacific islands and contaminating drinking water on others. The rising sea has endangered sacred sites and drowned some small islands near Kiribati and Tuvalu, including the islet of Tebua Tarawa, once a landmark for Tuvalu fishermen. Kiribati already has moved some roads inland on its main island as the rising Pacific Ocean eats into its shores. Rising sea levels are seeping into soils on some islands, as water tables rise. Soils in some areas are becoming too salty for growing most vegetables. In Tuvalu, according to a dispatch from Reuters, farmers "are beginning to grow their taro crops in tin containers filled with compost instead of traditional pits". Soil contamination is also an issue in the Bahamas, where fear has been expressed that the limestone that underlies the soil on many islands will absorb saline ocean water like a sponge.

Given expectations that global temperatures will rise further in coming decades, the sea level rise of the twentieth century is expected to be dwarfed by the rise during the twenty-first century. Scientists who attend to this issue typically estimate sea level rise possibilities in wide ranges, because no one knows how much ice will melt given a certain rise in temperatures over the poles. A team led by J.J. Wells Hoffman, for example, in 1983 estimated that sea levels in the year 2100 would be between 58 and 368 centimeters higher than they were in the 1980s. R. Thomas, in 1986, put the estimated range at 56 to 345 centimeters. The estimates vary by a factor of roughly seven, an indication of the uncertainty which plagues forecasts of how much the sea will rise under various assumptions about global warming.

Holocaust for Coral Reefs

Coral reefs, nurtured by an aquatic environment, are among the most productive, diverse ecosystems on earth. Sometimes called the rainforests of the sea, coral reefs cover much less than one percent of the world's ocean floors, while at the same time hosting more than a third of the marine species presently described by science, with many species remaining undocumented. Some of these organisms may provide new sources of anticancer compounds and other medicines. Coral reefs also protect shorelines from erosion by acting as break-waters which, if healthy, can repair themselves.

Coral reefs are extremely intolerant of small rises in water temperature, which destroy the ecosystem of the reef, leaving only a lifeless, bleached exoskeleton. By the early 1990s, an increasing number of the earth's coral reefs were turning white, a sign that they have been killed because of excessive heat. Most corals die at levels above 30 to 32 degrees C. Bleaching results when symbiotic zooxanthellae algae are expelled from coral reefs. Once bleached, the coral does not receive adequate nutrients or oxygen. If temperatures fall, the reefs may recover, but if thermal stress continues a large part of any given reef will die.

Pollution, disease, ultraviolet radiation, too much shade, and changes in the ocean's salinity also may contribute to coral bleaching. Warming seas are one among many human-induced changes that threaten coral reefs. Other problems include over-fishing, coastal development, nutrient runoff from agriculture and sewage, and sedimentation from logging on streams which feed into coastal waters. By one estimate, 58 percent of the world's coral reefs were being threatened by human activity by the late 1990s.

A scientific team led by Thomas J. Goreau plotted water temperatures at a number of locations in the tropical western Atlantic and Caribbean, finding slow rises beginning during the 1980s in such places as the Bahamas and Jamaica. In some cases, these temperature increases have been significant enough to dramatically accelerate the bleaching of coral reefs. The rate of bleaching correlates to general air temperature rises detected by global surveys of temperatures on land. "The fact that the high water temperatures of the late 1980s and mass coral bleaching have not been seen before in the Greater Caribbean during nearly 40 years of continuous study of coral reef ecosystems implies that temperatures have only recently exceeded tolerance limits of the corals". The authors of this paper cite similar results in Hawaiian waters. This article includes an extensive list of references which detail exactly what is happening to coral reefs around the world.

Some areas, including the Cook Islands and several locations in the Philippines, have been afflicted with massive coral bleaching since 1983. In 1995, warming in the Caribbean produced coral bleaching for the first time in Belize, as sea surface temperatures surpassed 29 degrees C (84 degrees F.). In 1997, Caribbean sea-surface temperatures reached 34 degrees C. (93 degrees F.) off southern Belize, and coral bleaching was accompanied by the deaths of many starfish and other sea life.

The central barrier reef near Belize suffered a mass coral die-off following record-high water temperatures in 1998. According to Richard Aronson, a marine biologist at the Dauphin Island Sea Laboratory in Alabama, this die-off represents the first complete collapse of a coral reef in the Caribbean Sea from bleaching. Richard Aronson, working with colleagues at the Smithsonian Institution in Washington D.C., said in Nature that lettuce coral was the most abundant species living in the area until 1998. Later surveys showed that "virtually all living colonies had been bleached white at almost every depth investigated". Core samples from the corals indicate that such a mass bleaching is unprecedented for at least the last 3,000 years. The research team found, "Complete bleaching was also evident in almost all species down to the lagoon floor at 21 meters".

By the late 1990s, coral bleaching had become epidemic in many tropical seas and oceans around the world. A World Wildlife Fund report found that massive coral bleaching occurred during the 1990s, in response to unusually high water temperatures, particularly during the El Nino years of 1997 and 1998. Sites of large-scale coral mortality included the Pacific Ocean, Indian Ocean, Red Sea, Persian Gulf, as well as the Mediterranean and Caribbean seas.

With more warming anticipated, the extent and severity of damage to coral reefs worldwide is expected to intensify. An Australian marine scientist, Dr. Terry Done, told Murray Hogarth of the Sydney Morning Herald that he regards coral reefs as the "canaries in the coalmine". During 1998, large coral reefs up to 1,000 years old died; and in some parts of the world up to 90 percent of reefs have been devastated, Done said.

The World Wildlife Fund sketches the worldwide nature of coral bleaching caused by rising ocean temperatures:

According to NOAA, coral bleaching was reported throughout the Indian Ocean and Caribbean in 1998, and throughout the Pacific, including Mexico, Panama, Galapagos, Papua New Guinea, American Samoa, and Australia's Great Barrier Reef starting in 1997.The severity and extent of coral bleaching in 1997–98 was widely acknowledged among coral reef scientists as unprecedented in recorded history.

The International Society for Reef Studies concluded that 1997 and 1998 had witnessed the most geographically widespread coral bleaching in the recorded histories of at least 32 countries and island nations. Reports of bleaching came from sites in all the major tropical oceans of the world, some for the first time in recorded history. The 1997–1998 bleaching episode was exceptionally severe, as a large number of corals died. According to one study, parts of Australia's Great Barrier Reef have been so severely affected that many of the usually robust corals, including one dated over 700 years of age, were badly damaged or had died during 1997 and 1998.

A World Wildlife Fund report indicates that corals are directly harmed not only by rising water temperatures, but also by increasing levels of atmospheric carbon dioxide.

Living coral reefs are composed of great numbers of coral animals covering a rigid skeleton formed by coral secretions of calcium carbonate. High levels of atmospheric CO2 alter water chemistry and reduce the calcification rate, and hence density, of coral skeletons. Some scientists believe that calcification probably already has decreased on some reefs, and predict that calcification could decrease 17 to 35 percent from pre industrial levels by 2100.

Coral reefs that suffer from reduced density due to decreased calcification face erosion by storm-tossed seas.

An international team of more than 250 marine scientists warned that conditions during 1998, the warmest year in recorded history on land and sea, had produced a mass breaching of many coral reefs around the world.

Scientists meeting in Townsville, near Australia's Great Barrier Reef, during November of 1998 declared that sea temperature rises predicted during the next fifty years pose a serious threat to nearly all of the world's coral reefs. Global coral bleaching and die-off was unprecedented in 1998 in geographic extent, depth and severity, said the International Coral Reef Initiative, an association of scientists and marine-park managers. The only major reef region spared from coral bleaching appears to be the central Pacific, the scientists said in a report on the status of the world's reefs.

Some of the most severe coral bleaching took place in the tropical Indian Ocean, where water temperatures during the 1998 El Nino reportedly rose to between three and five degrees C. above long-term averages in some areas. According to Clive Wilkinson and colleagues writing in the Swedish environmental journal Ambio,

Massive mortality occurred on the reefs of Sri Lanka, [the] Maldives, India, Kenya, Tanzania, and [the] Seychelles, with mortalities of up to 90 percent in many shallow areas. Coral death during 1998 was unprecedented in severity. Coral reefs of the Indian Ocean may prove to be an important signal of the potential effects of global climate change, and we should heed that warning.

According to another observer, "In some parts of the Indian Ocean reefs in the Maldives, Sri Lanka, Kenya and Tanzania were devastated with shallow reefs looking like graveyards".

In the Asia-Pacific region, the worst-hit coral reefs were near Japan, Taiwan, the Philippines, Vietnam, Thailand, Singapore, Indonesia and the islands of Palau. A scientists' statement, released at the same time in the United States and Australia, relied heavily on new satellite data from America's National Oceanic and Atmospheric Administration which indicated a degree of warming that they had not anticipated, especially in the tropics.

During May 1998, marine scientists said Australia's Great Barrier Reef (the world's largest living reef, stretching 1,300 miles), was experiencing its worst case of coral bleaching in recorded history. Australian marine scientists said bleaching had hit more than 60 percent of its 3,000 coral reefs and aerial surveys showed that 88 percent of inshore reefs were bleached, with 25 percent severely bleached.

By the year 2000 some coral reefs in the Indian and Pacific oceans had recovered somewhat from the rapid bleaching of 1998. Scientific studies indicated that some young corals in these areas had survived bleaching, and were helping to begin rebuilding some of the reefs. Terry Done, a senior research scientist at the Australian Institute of Marine Science, said that reefs may be "more resilient than we had thought." According to a report in Science, Done said that many corals may "not be able to mature and recover from the repeated bleaching forecast to accompany global warming". Done is best acquainted with some of the Indian Ocean corals which were characterized as looking like graveyards after the 1998 bleaching. He noted that most coral reefs require at least a decade to recover from bleaching. "This recovery won't do the reefs much good," Done said. "They'll no sooner get one or two years old before they'll be wiped out again".

Corals are not the only living things imperiled by rising ocean temperatures. C.D. Harvell and associates reported in Science, "In the past few decades, there has been a world-wide increase in the reports of diseases affecting marine organisms". This rapid increase in disease-related mortality among plants and animals of the seas and oceans is related by Harvell and associates to global warming, which acts to expand the range of many diseases, as well as human pollution and other forms of habitat degradation. Many of the diseases surveyed by these researchers are new to science. "The current trend toward a warming climate could result in modifications of many marine organisms' basic biological properties, thereby making them more susceptible to disease," Harvell and associates assert. One of the team's primary examples of warming 's synergy with disease is the mass bleaching of tropical corals worldwide during the 1997–1998 ENSO (El Nino Southern Oscillation). The authors of this article found that, along with the heat, warmer water often promoted diseases, which weakened the coral: "Demise of some corals is likely to have been accelerated by opportunistic infections". Harvell and colleagues call corals an "indicator species of a heightened disease load" throughout the oceans. The same authors also cite increased mortality of oysters in Chesapeake Bay, which they ascribe to warmer winters which have "decreased parasite mortality, resulting in oysters retaining heavy infections".

Warming and Decline of Life in the Ocean

Warming temperatures will force adaptation or extinction of many plant and animal species in the world's oceans. Great sea turtles, for example, will come ashore at their usual sites to lay their eggs on sand and rocks which have grown warmer. The temperature determines the sex ratio of the turtles, with female births declining as the environment warms. Scientists do not yet know how much warmth will be required before the last sea turtle mother will die. According to the World Commission on Environment and Development, one plant or animal species is going extinct somewhere in the world every minute, due mainly to human encroachment on (and pollution of) natural habitats.

Warming seas distort marine ecosystems in other ways. According to a report by Climate Solutions of Olympia Washington, a dramatic ocean-temperature increase near North America's West Coast began about 1977. Warming sea surface temperatures may interfere with phytoplankton production, with impacts rippling through the food web. Cooler, upwelling ocean water will break through warm surface waters less frequently, reducing nutrients available for plants and animals living in the oceans. By the 1990s, such decreases in productivity were detected near the California coast, where scientists have documented a measurable decrease in the abundance of zooplankton, the second level in the food web. By the 1990s, the abundance of zooplankton was 70 percent lower than it had been during the 1950s.

The decline of zooplankton raises fears for the survival of several fish species which feed on it. Water temperatures in some areas near the United States West Coast have increased one to two degrees C. during the last four decades. Ocean seabirds in the California Current have declined 90 percent since 1987, one indication of food web collapse. Reduced primary phytoplankton production usually means less overall fertility in marine ecosystems, including reduced fisheries, according to a report compiled by the World Wildlife Fund.

Harvard University epidemiologist Paul Epstein warns that rising temperatures in many ocean areas may be related to increasing outbreaks of algal toxins, bacteria, and viruses which affect large numbers of marine species, as well as shorebirds and mammals. "Of great concern," writes Epstein, "are diseases that attack coral and sea grasses, essential habitats that sustain mobile aquatic species". During the 1997–1998 El Nino, weak winds and strong sunlight created highly stratified, low-nutrient surface waters in the eastern Bering Sea, resulting in a massive bloom of coccolithophores, a type of phytoplankton usually associated with low-nutrient areas. According to the World Wildlife Fund, such a large-scale bloom had never before been documented in that area.

Scientists also have documented decreased reproduction and increased mortality in seabirds and marine-mammal populations in warming water. The World Wildlife Fund reports that Sooty Shearwaters off the California coast declined 90 percent during the late 1980s and early 1990s, as Cassin 's Auklets declined 50 percent. Zooplankton populations declined markedly at the same time. In Alaska, a severe decline in Shearwaters from 1997 to 1998 "was clearly due to starvation," according to the World Wildlife Fund.

By the late 1990s, the Mediterranean Sea had become home to at least 110 species of tropical fish for the first time in recorded history. The tropical fish, some of which are crowding out native species, have followed warming water into the Mediterranean through the Suez Canal and the Strait of Gibraltar. Barracuda and other tropical fish are migrating to the Mediterranean as water temperatures warm. "Eighty-five species previously unknown on a significant scale in the waters of Italy or the Levant have set up home there over the past ten years, according to research carried out by Professor Franco Andolaro and his team at the Palermo office of the Italian Marine Research Institute. They are competing for food with 550 indigenous species".

"What is in danger," Professor Andolaro said, "is the ecological balance of the Mediterranean basin. Tropical species now make up almost 20 percent of the fauna and the colonisation is continuing constantly". The Italian Environment Ministry says that water temperatures in the Mediterranean have increased about two degrees C. since 1970. Fifty-five of the new species are from the Red Sea and the remainder from the Atlantic. Ten of the Red Sea migrants are so numerous that they are now being fished commercially in the Mediterranean, including gold-band goatfish, striped-fin goatfish, Haifa grouper, streamlined spinefoot, the wahoo (a small tuna), banded barracuda, and Brazilian lizard fish.

News of the influx has caused considerable excitement in Italy, where tourism to exotic destinations is a burgeoning business. "We won't have to go as far as the Seychelles, Sharm el-Sheikh or the Maldives to enjoy the most fascinating marine beauty," Il Giornale of Milan commented. "Today we can admire these curious and highly coloured tropical fish by simply taking a swim in Liguria or Sardinia".

Another potential problem with warming seas is an expected reduction in the pH factor, indicating an increase in relative acidity of the water. Wilfrid Bach writes,

A CO2 doubling in the atmosphere might reduce the pH of surface sea water from the usual 8.1 to 7.6. It has been shown that a reduction of the pH to 7.6 would increase the copper-ion activity by ten-fold, a change to which the marine phytoplankton would react dramatically a pH of 7.7 would cause the death of fish larvae.

Rising sea levels have been killing trees in some of Florida's coastal parks and nature reserves, one indication of the fact that the world's oceans have been rising for about 10,000 years, since the last glacial maximum, producing gradual saltwater encroachment into relatively flat coastal land areas. In some areas, certain species of trees do not die, but do cease to reproduce, causing them to expire as a species within one generation. Red cedars are among the first species of tree to die, while cabbage palms survive saltwater intrusion longer than most.

Dimensions of Sea Level Rise

In It's a Matter of Survival, Anita Gordon and David Suzuki quote Stephen Leatherman regarding an anticipated worldwide sea level rise of 1.5 meters (five feet) during the twenty-first century: "We're looking at land loss greater than we've seen at any time in human history. There will be so much total immersion that maps will have to be redrawn along many of the low-lying coastal areas, such as along the U.S. Atlantic and Gulf coasts".

Anticipated global warming will raise worldwide sea level not only because of melting ice, but also because the molecular structure of liquid water expands as it is heated. The same amount of liquid water will occupy more volume as it warms. James G. Titus and Vijay Narayanan estimate that about 20 centimeters of the next century's sea level rise will be caused by thermal expansion of existing waters. The amount of expansion depends ultimately on how deeply the warmer surface waters mix with lower levels.

"Although global temperatures are projected to rise 20 percent less during the twenty-second century than in the twenty-first, thermal expansion is likely to be 20 to 40 percent more, due to the delayed response of expansion to higher temperatures".

Projections of sea level rise are plagued with a large degree of uncertainty, in large part because no one knows how much climatic "forcing" will be necessary before major ice masses, such as the West Antarctic Ice Sheet, begin to contribute significantly to world sea level. A report by the Department of Energy estimated that over a period of two to five hundred years, the West Antarctic Ice Sheet could disintegrate, raising sea level six meters (20 feet). Studies after 1983, however, have focused on the next century. Most recent estimates of possible sea level rise due to global warming generally fall in a range of 50 to 200 cm (two to seven feet) by 2100. Studies since 1990, however, generally suggest that a 50 to 200 centimeter rise in mean sea level is more likely to take 150 to 200 years.

Global sea level was about 328 feet (100 meters) lower at the peak of the last ice age. Since then, a little more than half the Earth's glacial-maximum ice mass has melted; if the rest were to melt, the sea level would rise an estimated 280 feet from present levels. The earth last witnessed such a state for a sustained period during the age of the dinosaurs, about 65 million years ago, when the planet had no permanent ice, and the ambient global temperature was as much as 20 degrees F. higher than today. Very probably following a cataclysmic collision with an asteroid or comet, the Earth entered a period during which dust obscured the sun, causing notable cooling, and giving rise to the new cycle of ice ages alternating with warmer interglacial periods which has characterized the last several million years. During this period, warm-blooded animals (including Homo sapiens) have become the Earth's dominant species.

Even a minor rise in sea level would inundate wetlands and lowlands, accelerate coastal erosion, exacerbate coastal flooding, threaten coastal structures, raise water tables, and increase the salinity of rivers, bays, and aquifers. Most of the wetlands and lowlands of the United States are located along the Gulf of Mexico and Atlantic coasts south of central New Jersey. Shoreline erosion is not limited to the East and Gulf coasts. About 950 miles (86 percent) of California's 1,100-mile coastline eroded during the late twentieth century. Large mudflats also ring parts of Puget Sound, and parts of Washington State's ocean shoreline also are relatively flat.

An Environmental Protection Agency report issued during 1997 said that a one-meter rise in sea level will inundate about 7000 square miles of dry land in the United States, including between 50 and 80 percent of U.S. wetlands. The wetlands, the coastal interface between fresh and salt water, are important to the breeding of many fish species which are particularly sensitive to water temperature, acidity, and salinity. The wetlands of Louisiana, which once hosted stands of oak and cypress, are now losing them at a rate of 55 square miles a year as rising ocean waters increase the salinity of swamplands. In some cases, the increase in salinity is being worsened by construction of canals by oil companies.

The State of South Carolina has enacted a Beachfront Management Act, which curtails shorefront development. The State of Maine also limits development in any area that would be eroded by a 90-centimeter rise in average sea level. Land reclamation projects in San Francisco and Hong Kong now include a safety margin for future sea level rise, as do new seawalls in eastern Britain and the Netherlands.

The EPA has compiled reports projecting the costs of protecting ocean-resort communities by pumping sand onto beaches and gradually raising barrier islands as seas rise. The EPA also estimated the costs of protecting developed areas along sheltered waters through the use of levees (dikes) and bulkheads, as well as the anticipated loss of coastal wetlands and undeveloped lowlands. The total cost to mitigate a one-meter sea level rise would be $270 to $475 billion, according to EPA estimates. These reports generally assume the efficacy of human artifice, often ignoring the fact that seawalls built to protect coastal developments from erosion may actually hasten its speed because they interfere with the natural seasonal cycle of sand replenishment.

The net effect of sea level rise in any particular place must be adjusted for the gradual rise or fall of the land itself. Some areas that were covered with ice during the last glacial maximum are rising. An example is Stockholm, Sweden. Parts of Canada and Scotland also are slowly rising following the melting of glaciers several thousand years ago, while much of the United States' Atlantic Coast has subsided about a foot during the twentieth century. Along a relatively level, sandy shoreline, such a change may cause a beach to lose as much as 100 feet. Many Atlantic beaches have been receding as much as three feet per year; in some areas of the Gulf of Mexico coast, the rate averages five feet per year.

Coastal Louisiana is subsiding 0.4 inch per year, an unusually rapid rate, due to removal of oil, gas, and groundwater. Because it is subsiding so quickly, the Mississippi River Delta in Louisiana is probably the area of the United States that is most vulnerable to sea level rises caused by global warming. Land loss in Louisiana due to subsidence and rising waters has been estimated at about one million acres during the twentieth century; by 1990, roughly 50 square miles a year were being lost. The highest point in the city of New Orleans is only 13 feet above mean sea level, as the ground under the city sinks three feet per century.

As sea levels rise, cities along the U.S. East and Gulf coasts (and other coastal locales around the world) may find saline water seeping into their drinking-water supplies. Cities most at risk in the United States may be New York City and Philadelphia. In the Philadelphia area, a sea level rise of a third of a meter would require a 12 percent rise in reservoir capacity to prevent saltwater intrusion into system intakes on the Delaware River. Other cities around the world that lie near seacoasts also may face similar problems: Amsterdam, Rotterdam, Liverpool, Istanbul, Venice, Barcelona, Gothenburg, St. Petersburg, Calcutta, among others.

New York City, Philadelphia, and much of California's Central Valley get their water from areas that are just upstream from areas in which the water is salty during droughts. Farmers in central New Jersey as well as the city of Camden rely on the Potomac-Raritan-Magothy aquifer, which could become salty if sea level rises. The South Florida Water Management District already spends millions of dollars per year to prevent Miami's Biscayne aquifer from becoming salty. A rise in sea level also would enable salt water to penetrate further inland, as well as upstream into rivers, bays, wetlands, and aquifers, which would be harmful to some aquatic plants and animals, and would threaten human uses of water. Increased salinity probably will threaten oyster harvests in the Delaware and Chesapeake bays.

Along the United States Eastern Seaboard, oceanfront property, some of the most valuable real estate in the country, may be vulnerable to sea level rises caused by global warming. "In Massachusetts," writes Bill McKibben, "between three thousand and ten thousand acres of ocean-front land worth between $3 billion and $10 billion might disappear by 2025, and that figure does not include land lost to growing ponds and bogs created as the rising sea lifts the water table".

Chesapeake Bay "is subject to all the SLR [sea-level rise] impacts: erosion, inundation of low-lying lands and wetlands loss, salt-water intrusion into aquifers and surface waters, higher water tables, and increased flooding and storm damage". Crab populations that rely on aquatic vegetation for protection during their early life stages are declining in Chesapeake Bay. In the same area, many small islands, which have been important rookeries for several species of birds, are eroding. The Cape Hatteras lighthouse, first built in 1803 (then, as now, the tallest lighthouse in the United States), is being moved inland. It was 1,600 feet from shore when built, but 120 feet from open water when it was moved this year.

Texas may face special hazards from sea level rise along the Gulf Coast. In the modulated tones of science, a report outlining possible outcomes of the greenhouse effect on Texas sketches severe problems with land subsidence and sea level rise along parts of the state's coast:

The Houston-Galveston urban region, with its high groundwater table, subsidence, and long history of severe flooding along its coastline and bayous, will be particularly vulnerable, and could experience permanent loss of urban land in sensitive areas, necessitating extensive relocation programs.

In addition, a rise in sea level may imperil drinking water supplies in the coastal regions of Texas, the state's fastest-growing region in terms of human population and development. The Houston-Galveston area is additionally vulnerable because the land is relatively flat for several miles inland, and has been experiencing subsidence (sinking) as groundwater has been removed for human consumption. The cities of Beaumont, Port Arthur, Orange, Freeport, and Corpus Christi subsided as much as a foot between 1906 and 1974. Some small areas have sunk as much as ten feet during that period. Subsidence rates could increase as more groundwater is withdrawn.

Coastal erosion is an additional problem in the Houston-Galveston area. At Sargent Beach, the shoreline eroded about 1,000 feet between 1956 and the early 1990s. Galveston built a seawall to protect against oceanic flooding during the early 1900s, following a disastrous hurricane and storm surge there in 1900. When the seawall was constructed, it faced a beach that averaged about 300 feet wide. During ensuing years, the beach has eroded; by about 1940, more rocks were required to protect the original seawall. Most of these rocks sank during the following few years, requiring even more rocks.

The EPA estimates that if no measures are taken to hold back the sea, a one-meter rise in sea level will inundate 14,000 square miles of coastal land area in the United States with salt water during the twenty-first century. Roughly 1500 square kilometers (600–700 square miles) of densely developed coastal lowlands could be protected at a cost of approximately $1,000 to $2,000 per year for a typical coastal housing lot. Given high coastal property values, holding back the sea could be cost effective, the report contends. The report adds, however, that the environmental consequences of holding back the sea may be unacceptable. Although the most common engineering solution for protecting the ocean coast pumping sand would allow beaches, levees, and bulkheads to be maintained along sheltered waters, the same actions would gradually eliminate most of the nation's wetlands. Based on this probability, the report finds, "To ensure the long-term survival of coastal wetlands, federal and state environmental agencies should begin to lay the groundwork for a gradual abandonment of coastal lowlands [by human inhabitants] as sea level rises". Barrier islands and sand-spits of land along the Atlantic and Gulf coasts are among the most vulnerable areas to rises in sea level. Most coastal barrier islands are long, narrow spits of sand with ocean on one side and a bay on the other. Typically, the oceanfront shore of any given island usually ranges from two to four meters above high tide, while the bay side is less than a meter above high water. Thus, even a one-meter rise in sea level would threaten much of these lands (and their ranks of hotels, businesses, and homes) with saltwater inundation. Erosion, moreover, threatens the high parts of these islands, and is generally viewed as a more immediate problem than the inundation of barrier islands' bay sides. A rise in sea level can cause an ocean beach to retreat by considerably more than the retreat due to inundation alone, according to an EPA report published in 1997:

The shape of a beach profile is determined by the pattern of waves striking the shore; generally, the visible part of the beach is much steeper than the underwater portion that comprises most of the active "surf zone." While inundation alone is determined by the slope of the land just above the water, Bruun showed that the total shoreline retreat from a rise in sea level depends on the average slope of the entire beach profile.

Studies suggest that a one-meter rise in sea level would generally cause beaches to erode 50 to 100 meters from the New England to Maryland coasts, about 200 meters along the Carolina coasts, 100 to 1000 meters along the Florida coast, and 200 to 400 meters along the California coast. Because most U.S. recreational beaches are less than 30 meters (100 feet) wide at high tide, even a 30 centimeter (one foot) rise in sea level could cause serious loses to large areas of high-priced coastal real estate. Many coastal areas are becoming more vulnerable to flooding for four reasons, according to the 1997 EPA report:

(1) A higher sea level provides a higher base for storm surges to build upon; a onemeter rise in sea level would thus enable a 15-year storm to flood many areas that today are only flooded by a 100-year storm.

(2) Beach erosion would leave particular properties more vulnerable to storm waves.

(3) Higher water levels would increase flooding due to rainstorms by reducing coastal drainage.

(4) A rise in sea level would raise water tables.

Possible responses to anticipated sea level rise generally fall into three categories, according to the EPA: erection of walls to hold back the sea, allowing the sea to advance and adapting to it, and raising the land itself. For more than five centuries, the Dutch have used dikes and windmills to prevent inundation from the North Sea. By contrast, some English towns have been rebuilt landward as structures and land were lost to erosion; the town of Dunwich, England, has had to rebuild its church seven times in the last seven centuries. More recently, rapidly subsiding communities such as Galveston, Texas, have used fill to raise land elevations; The U.S. Army Corps of Engineers and coastal states regularly pump sand from offshore to counteract beach erosion. Venice, Italy, has used all three responses: allowing the sea to advance into the canals, while raising some lowlands, and erecting storm protection barriers.

Global Warming and Tropical Cyclones

Kerry Emanuel, a hurricane specialist at the Massachusetts Institute of Technology, published an article in the Journal of the Atmospheric Sciences in which he argued that hurricane intensity is governed, in part, by the degree of thermodynamic disequilibrium between the atmosphere and the underlying ocean. Therefore, Emanuel reasoned, warmer ocean waters would breed more intense tropical cyclones. By the year 2000, Emanuel's forecast of "hypercanes" had become gist for at least one best-selling book, The Coming Superstorm by Art Bell and Whitley Strieber, which rose to number 15 on the New York Times best-seller list for nonfiction hardbound books during January 2000.

In 1987, Emanuel published an article in Nature asserting that a three degree C. increase in sea surface temperature could increase the potential destructive power (as measured by the square of the wind speed) of storms by 40 to 50 percent. Emanuel asserted in 1988 that a warming of the sea surface by 6 to 10 degrees C. (assuming no temperature change in the lower stratosphere) would make a supersized, ultrapowerful "hypercane" theoretically possible (Emanuel, Maximum Intensity, 1988; Toward a General Theory, 1988). Emanuel's theory has sparked intense debate among hurricane specialists. Critics of his theory responded that even the most pessimistic climate models do not project such a large amount of warming in the tropics.

The temperature of the surface water over which a hurricane passes is important to its growth and development, temperature is only one of several variables that influence the career of any given storm. Emanuel, writing in Nature, outlines the major factors influencing hurricane intensity: "the storm's initial intensity, the thermodynamic state of the atmosphere through which it moves, and the heat exchange with the upper layer of the ocean under the core of the hurricane". While climate modelers have had some success forecasting the tracks of hurricanes, intensity is a much tougher theoretical nut to crack. Within a few hours, a hurricane such as Opal may intensify from a category 2, with strongest winds about 100 miles an hour, to a devastating category 5, with winds above 150 m.p.h. Opal's intensification was due mainly to its passage over a patch of unusually warm water in the Gulf of Mexico, together with upper-atmospheric dynamics which were favorable for storm development. The speed and complexity of hurricane intensity in the very short term has thus far eluded climate modelers.

The IPCC 's First Assessment was rather ambivalent regarding global warming 's possible effects on hurricanes. The IPCC report said, "There is some evidence from model simulations and empirical considerations that the frequency per year, intensity, and area of disturbance of tropical cyclones may increase, though it is not yet compelling".

A study by T.R. Knutson, R.E. Tuleya, and Y. Kurihara used a climate-simulation model to study 51 Pacific Ocean cyclones under present day conditions and under simulated conditions which include higher levels of carbon dioxide. They found that a warming of sea surface temperatures in the Western Pacific of about 2.2 degrees C. would increase the storms' wind speeds an average of 5 to 12 percent, and decrease their central pressures 7 to 20 millibars. The study suggests that global warming will produce more destructive tropical cyclones and that people living in the possible paths of these storms are courting potential disaster.

A report by Christopher W. Landsea of the National Oceanic and Atmospheric Administration's Miami, Florida, office argues that an increase in carbon dioxide levels may raise the threshold temperature in which hurricanes thrive (presently above 80 degrees F.) in proportion to the rise in temperature. This change might nullify the increase in strength that is implied by warmer water surface temperatures. Land sea also speculates that increased frequency of El Nino-type weather patterns in the Pacific tend to dampen hurricane development in the Atlantic, although storm frequency and intensity often rises under El Nino conditions in the eastern Pacific.

Landsea contends that intensity of Atlantic hurricanes has decreased since the middle of the twentieth century. Landsea and associates also assert that hurricane frequency and intensity had not increased during the past half-century. Landsea and colleagues wrote that "a long-term (five decade) downward trend continues to be evident primarily in the frequency of intense hurricanes. In addition, the mean maximum intensity (i.e., averaged over all cyclones in a season) has decreased".

Landsea believes that any warming-induced change in hurricane frequency and intensity will probably be lost in the "noise" of year-to-year hurricane variability. He concludes, "Overall, these suggested changes are quite small compared to the observed large natural variability of hurricanes, typhoons and tropical cyclones. However, more study is needed to better understand the complex interaction between these storms and the tropical atmosphere and ocean".

Three greenhouse skeptics have challenged Emanuel's theories. They assembled hurricane data for the central Atlantic, the East Coast of the United States, the Gulf of Mexico, and the Caribbean Sea between 1947 to 1987, comparing their information to estimates of the sea surface temperatures in the Northern Hemisphere. Idso, Balling, and Cerveny concluded, "There is basically no trend of any sort in the number of hurricanes experienced in any of the four regions with respect to variations in temperature".

Doubts also have been raised about the relationship of warming seas and intensity of tropical cyclones by scientists whose views are less politicized than those of Idso, Balling, and Cerveny. A.J. Broccoli and S. Manabe included cloud-related feedbacks in their models and found a 15 percent reduction in the number of days with hurricanes, even assuming a temperature rise caused by a doubling of atmospheric carbon dioxide. Broccoli and Manabe noted that their results were easily skewed by (and highly dependent upon) how they decided to represent cloud processes within their models.

On the other hand, S.T. O'Brien and colleagues assert that doubling the level of atmospheric carbon dioxide (increasing tropical sea-surface temperatures between one and four degrees C.) will double the number of hurricanes, and increase their strength by 40 to 60 percent. O'Brien and associates also believe that warmer average temperatures will extend the hurricane season, because the season effectively ends for any given area when water temperature falls below 26 degrees C. (80 degrees F.) R.J. Haarsma and colleagues also estimate that a doubling of greenhouse-gases levels will increase the frequency of hurricanes by 50 percent and increase the average intensity of the storms by 20 percent.

J. Lighthill and associates argue that while global warming may exert some influence on cyclone formation in the tropics, natural variability is more important. This position drew a reply from Emanuel. L. Bengtsson, a member of the German Max Planck Institut für Meteorologie, contends that global warming will strengthen the upper-level westerlies in areas where hurricanes usually develop, inhibiting storm development and intensity, negating any boost the storms may get from warmer sea-surface temperatures.

The IPCC, in its Second Assessment, restated its earlier position that the state of science on the subject does not permit a conclusion regarding whether global warming will affect the number and intensity of tropical cyclones. Tom Karl and colleagues wrote two years later in Scientific American, "Overall, it seems unlikely that tropical cyclones will increase significantly on a global scale. In some regions, activity may escalate; in others, it will lessen".

Global Warming and Ocean Circulation

Circulation patterns within the oceans carry water around the world at several levels, mixing it laterally as well as vertically. Without such circulation, deep waters would become depleted of oxygen and most marine life there would suffocate. A report by the World Wildlife Fund suggests that changes in the circulation of the ocean "are likely to be affected by global climate change, and in turn are likely to affect future climate change". Scientific attention has focused on the "thermohaline circulation" of the North Atlantic as one major example of ocean-atmosphere interaction which could shape climate change on nearby land masses, especially Western Europe.

Because thermohaline circulation is driven by density differences of water masses, it is extremely sensitive to influxes of fresh water of the type which are expected to occur with global warming due to melting sea ice and increased precipitation. Fresh water makes ocean water less saline and therefore less dense, so it doesn't sink as quickly. If enough fresh water is added, deep-water formation in the North Atlantic may cease altogether. Relatively small amounts may be sufficient to alter thermohaline circulation; some climate models suggest that, under "business-as-usual" conditions, a complete shutdown of thermohaline circulation in the Atlantic could eventually occur.

Regarding the relationship between thermohaline circulation and the climate of Europe, Giancarlo G. Bianchi and I. Nicholas McCave, writing in Nature, assert, The main concern for future climate must be that a possible increase in melting of the Greenland ice sheet resulting from anthropogenically induced atmospheric warming may reach a critical level where the "conveyor belt" will flip to its early Holocene operational mode. The resulting perturbations could conceivably result in climate extremes exceeding those of the Little Ice Age for northern Europe. Without such perturbations, the climate looks likely to warm for several hundred years.

Carsten Ruhlemann and colleagues wrote in Nature that the Earth's emergence from its last glaciation was punctuated by several short, sharp periods during which glacial conditions returned. These variations seem to have been triggered by the type of changes in North Atlantic thermohaline circulation which are considered likely in a world warmed by increasing emissions of greenhouse gases. "The thermohaline circulation was the important trigger for these rapid climate changes," they write. Climate changes which diminish thermohaline circulation tend to divert the Gulf Stream southward, cooling the North Atlantic Ocean (as well as Greenland and Western Europe), but "warming the western tropical North Atlantic and most of the Southern Hemisphere".

Between 1968 and 1972, an area which came to be known as "The Great Salinity Anomaly" was documented. The anomaly is a large pool of relatively fresh water (vis-à-vis other sea water), resulting from two years of unusually rapid Arctic sea ice melt. This flow of relatively fresh water shut down deep-water formation in the Labrador Sea, an important site for deep-water transport in the North Atlantic. The circulation necessary to transport deep water resumed when the fresher water was removed.

Such impairment of ocean circulation may compound atmospheric increases in greenhouse gases because it impedes the mixture of warmer surface water with cooler water in the depths of the ocean. The same mechanism would decrease the ocean's ability to absorb carbon dioxide from the atmosphere, in part because warmer ocean water has less absorption capacity than cooler water. "Additionally," according to a report by the World Wildlife Fund, "a slowing of deep-water formation in the Atlantic would likely reduce the transport of oceanic heat to the European continent. European cities along the Atlantic Sea-board may begin to cool, approaching the cooler temperatures of their latitudinal counterparts in the Pacific".

Scientific attention continues to be concentrated on the future of thermohaline circulation in the North Atlantic. Modeling work on North Atlantic ocean circulation reported in 1999 by R.A. Wood, A.B. Keen, J.F.B. Mitchell, and J.M. Gregory indicates "a dramatic change in the Atlantic occurring over the next few decades: a complete shutdown of one of the two main 'pumps' driving the formation of North Atlantic Deep Water, namely the one in the Labrador Sea". Britain's Hadley Climate Center projects that global warming will reduce overall thermohaline circulation by 25 percent, with the most notable effects in the area's southern reaches.

Collapse of the Labrador Current, which Richard A. Wood and colleagues believe is possible between the years 2000 and 2030, "could have serious consequences for marine ecosystems, including seabirds, as they depend not only on specific temperature conditions but also on nutrients supplied by oceanic mixing and currents". As of 2000, observations had not confirmed the predicted breakdown in the Labrador Current; it is believed that decay of the current is being forestalled "by the high phase of the North Atlantic Oscillation". Nevertheless, according to Stefan Rahmstorf, "The simulated ending of the Labrador Sea convection found in the Hadley Center model is perhaps the most convincing demonstration so far of [a] qualitative threshold being crossed because of global warming".

Changes in ocean circulation are not limited to the Atlantic. Northward flow of ocean water from the Pacific Ocean through the Bering Strait into the Arctic Ocean also appears to have slowed, according to the World Wildlife Fund report.

This change reduces northward flow of nutrient-rich North Pacific Ocean water over the Bering Sea shelf, "potentially reducing the overall primary productivity in the region".

Oceanographers have found similar trends in parts of the oceans bordering Antarctica. Marine geochemist Wallace Broecker of Columbia University's Lamont-Doherty Earth Observatory and colleagues have found, "The renewal of deep waters by sinking surface waters near Antarctica has slowed to only one-third of its flow a century or two ago". The work of Wallace Broecker, Stewart Sutherland, and Tsung-Hung Peng suggests that the slowing of deep-water formation may be related to warming attending the end of Europe's "Little Ice Age" (roughly 1400 to 1800 A.D.). Furthermore, they posit, "A see-sawing of deep water production between the northern Atlantic and the Southern oceans may lie at the heart of the 1,500-year ice-rafting cycle".

"This huge, climate-altering change in the oceans if it's real would greatly complicate attempts to understand how the ocean and climate are responding to the buildup of greenhouse gases in the atmosphere," writes Richard A. Kerr in Science. Kerr quoted Jorge Sarmiento, an ocean-circulation modeler at Princeton University, as saying that the work of Broecker and colleagues is "really interesting and provocative." Sarmiento also said, however, that he was "very uneasy about the calculations. I find the paper more of a stimulation to further work than what I would accept as proven fact".

Variations in the thermohaline circulation of the North Atlantic can have marked climatic consequences in Mesoamericam as well as in Europe. In The Great Maya Droughts, for example, Richardson B. Gill postulates that a southward shift in the Gulf Stream forced by cold deep-water flow from the north cooled climate in northwestern Europe between 800 and 1000 A.D. The same shift is said by Gill to have displaced the North Atlantic High southwest-ward, causing many years of severe drought in regions populated by the classic Maya civilization in Mesoamerica. Gill makes a case that these droughts played a major role in collapse of the classic Maya civilization.

As global temperatures have risen since 1980, the El Nino Southern Oscillation (ENSO) has occurred more frequently than at any other time in the century and a half that detailed worldwide records have been kept. Climate modeling by A. Timmermann and associates tends to support the idea that further warming will enhance the prominence of El Nino-type events in world oceanic and terrestrial circulation and climate patterns. The forecasts of Timmermann and associates come with an added twist: the La Nina side of this oscillation may be rarer, but possibly colder as well. In brief, their models indicate that weather may become generally wilder as well as warmer as levels of greenhouse gases increase. The researchers believe that in a warmer world climate will change in several ways.

The climatic effects will be threefold. First, the mean climate in the tropical Pacific region will change toward a state corresponding to present-day El Nino conditions. It is therefore likely that events typical of El Nino will also become more frequent. Second, a stronger inter-annual variability will be superimposed on the changes in the mean state, so year-to-year variations may become more extreme under enhanced greenhouse conditions. Third, the inter-annual variability will be more strongly skewed, with cold events (relative to the warmer mean state) becoming more frequent.

Global warming may be giving El Nino a "jump start" by providing ocean water at appropriate latitudes with increased warmth. The two most severe El Nino events since 1850 occurred late in the twentieth century the first in 1982–1983, the second in 1997–1998. The second severe El Nino caused 23,000 deaths and $33 billion worth of property damage to human infrastructure around the world.

Researchers have speculated that the strength of these two El Nino events may have been influenced not only by warmer ocean waters in general, but also by peak activity in the "interdecadal ENSO," another climate cycle in the eastern tropical Pacific in which "the temperature swings slowly from warm to cold and back over ten to twenty years" Yet another cycle in Pacific Ocean temperature (as yet unnamed), which involves the central reaches of the tropical Pacific Ocean, began to warm in the late 1970s and, like the "interdecadal ENSO," seems to have reinforced the better-known El Nino cycle from the 1980s into the 1990s. Another possible influence on the strength of El Nino events is the Pacific Decadal Oscillation (PDO), in the colder waters of the central northern Pacific. Researchers have proposed that a cooler-than-usual northern Pacific Ocean may help intensify storms spawned by El Nino conditions, especially in North America.

El Nino conditions do not always occur in warm weather, however. The lead author of an article in Science, Tammy Rittenour, said that traces of El Nino-like conditions have been found in the last ice age, 13,500 to 17,500 years ago. The work of Rittenour and colleagues suggests that temperature may be only one of several factors affecting the ebb and flow of several El Nino cycles. "The results fit well with new climate models that suggest that periods of weakened El Ninos rhythmically alternate with the current mode of strong El Ninos. Driving these swings, at least in the models, are periodic variations of solar heating as Earth wobbles on its spin axis".

Rittenour was surprised by indications that her climate research was indicating El Nino-type events during a glacial period. She had begun by investigating the rate at which melting drained glacial Lake Hitchcock, in present day southern New England. "It kind of threw me back," Rittenour told a reporter. "Coming up with El Nino was kind of a shock".

4

Sustainable Development: Kyoto Protocol and the Multilateral

Introduction

This chapter makes an initial attempt to assess the interaction between two, potentially revolutionary, international legal instruments, both of which attempt to influence the pattern of private sector investment from developed to developing countries. The first is the 1997 Kyoto Protocol to the United Nations Framework Convention on Climate Change, which, when in force, will stimulate investments in the developing world in projects that reduce emissions of greenhouse gases (GHGs) through a "clean development mechanism" (CDM). The second is the draft Multilateral Agreement on Investment (MAI). If adopted and ratified, the MAI will set high global standards for the protection of investors and investments against discrimination, and against illegal expropriation. Although negotiated under the auspices of the Organization for Economic Cooperation and Development (OECD), the MAI will be open to membership of both developed and developing countries.

This discussion is necessarily speculative, as the detailed investment rules of the Kyoto Protocol's CDM have yet to be agreed. The MAI remains in draft form, and its adoption is by no means assured. Nevertheless, a study of the two instruments, even in their present forms, is of interest as it allows an exploration of the potential conflicts and synergies that may exist between efforts to use international law both to promote and to channel international investment flows. Furthermore, the greater use of market mechanisms in multilateral environmental agreements and the powerful trend towards strengthened investor protection both indicate the advisability of exploring these issues, even if neither the Kyoto Protocol nor the MAI take quite the final shapes that we presume.

While both agreements aim to promote investment flows, the MAI attempts to limit the extent to which states intervene in the market-place in order to channel such investments towards domestic policy objectives. Under the MAI, domestic regulations may not either directly or indirectly discriminate between domestic and foreign investors or investments. The Kyoto Protocol's CDM, while described as a "market mechanism, " may in fact require extensive state intervention in investment decisions. The CDM does not view all categories of investor or investment equally, aiming instead to promote investments from developed into developing countries in activities that reduce GHG emissions.

The potential for conflict between economic liberalization regimes such as the MAI and the World Trade Organization (WTO) Agreements, and multilateral environmental agreements (MEAs) such as the climate-change regime, has generated volumes of academic literature. As yet, no formal disputes have arisen. There is, however, a marked trend in the regimes of MEAs to draw upon market mechanisms and market interventions to achieve policy goals. These include investment incentives, such as the CDM, tradeable allowances (also anticipated under the Kyoto Protocol), and technical barriers to trade arising from the prior informed-consent requirements under the emerging hazardous chemical and biosafety treaties.

Perceptions of the likelihood and the seriousness of the potential conflict differ widely. Hoping for synergy but seeking to avoid conflict, negotiators are increasingly looking for ways to carve out exceptions that will allow potentially competing rules to accommodate each other. However, as long as these regimes remain under the authority of distinct governance structures and dispute-settlement procedures, a risk persists that one set of objectives will have to concede to the other in an ad hoc and incoherent manner.

Clean Development Mechanism

Under Article 12 of the Kyoto Protocol, a CDM will promote investments in climate-friendly project activities in non-Annex I Parties. The CDM will be governed by a multilaterally agreed set of rules and will operate under the supervision of the Conference of the Parties serving as the meeting of the Parties to the Protocol (COP/MOP) and an executive board. Emissions reductions accruing from the "project activities" carried out by non-Annex I Parties, once certified under agreed principles, may be acquired and used by Annex I Parties to contribute to compliance with their emissions reductions obligations under Article 3 of the Protocol.

A well-designed CDM will have to offer positive incentives to the kind of investors and investments that will deliver "emission reductions that are additional to those that would occur in the absence" of the investment. Because CDM offsets will, in effect, allow Annex I Parties to increase their emissions above their Article 3 obligations, proof of CDM additionally is essential to ensure that overall global emissions are constrained.

Articles 12 and 3 of the Protocol leave open a number of variations on the steps in the CDM project activity cycle, i.e. the process whereby a project activity becomes a certified emissions reduction (CER) available for use as part of an Annex I Party's "assigned amount" under Article 3. It is possible to set out six essential steps in the CDM project activity cycle, as shown in table 4.1.

The simplest of potential CDM transactions would involve bilateral, project-by-project transactions between one investor and one host. These transactions may be purely intergovernmental in character or may also involve private investors and hosts.

More complex transactions are those proposed by a number of multilateral development banks and private financial service providers, which would draw together multiple investors and multiple projects, and multilateralize various steps in the CDM project activity cycle. Additional "service providers" may be required to pool eligible project activities into portfolios, to auction project activities or CERs to the highest bidder, or simply to match investors and project activities in a transparent manner.

Table 4.1. Basic steps in the CDM project activity cycle

Step	KP Article	Participant(s)
1. Approval of project activity	12.5(a)	S_a and S_b
2. Certification of environmental and financial additionality of project activity	12.5	Operational entity
3. Auditing and verification of emissions reductions	12.7	Independent auditor/ verifier
4. Assessment of proceeds for administrative and adaptation costs	12.8	To be determined by COP/MOP
5. Transfer/acquisition of CER	3.12	S_a and S_b
6. Addition of CER to assigned amount	3.12	S_e

More complex, multilateral arrangements are thought to achieve economies of scale, to increase the transparency and availability of information about transactions, to encourage the participation of smaller hosts and investors, to promote balanced bargaining positions between investors and hosts, and to spread the risk of project failure across a larger number of participants and project activities.

Multilateral Agreement on Investment

Drawing upon precedents from bilateral and plurilateral investment treaties, the MAI is intended to provide a "high standard" of protection to foreign investors and investments when operating overseas. The MAI would achieve this by constraining a very broad range of government regulatory activity that might directly or indirectly harm the interests of foreign investors or that might adversely affect the value of foreign-owned investments.

Host governments would be prohibited from applying discriminatory treatment against foreign investors or investors within their jurisdiction. National treatment (NT) and most-favoured nation (MFN) obligations within the MAI would require governments to extend to foreign investors and investments the same treatment that (or better treatment than) they extend to domestic investors and investments, and to treat foreign investors from different countries equally. The MAI would prohibit both de facto and de jure discrimination by the host state. This means that host country regulations that discriminate between domestic and foreign investors either expressly, or in their effect, would be open to challenge.

Parties to the MAI would also be prevented from forcing investors to conform to certain types of "performance requirements." Host governments would be prevented from conditioning the right to invest in, for example, the use by the investor of a certain percentage of local goods, or on the transfer of technology. This prohibition holds even if it is applied equally on domestic or foreign investors.

Foreign investors would also be protected from illegal expropriation of their investment assets and, in circumstances in which investments were legally expropriated, would be entitled to a high standard of compensation. Expropriation is denned broadly to include not just the direct taking of an investment but also indirect takings including "measures having equivalent effect." The breadth of this definition is intended to include so-called "creeping expropriation," whereby government taxation or regulations, short of a direct taking, may diminish the value of an investment.

The level of investment protection is further strengthened by the MAI's very broad definition of the types of "investment" that it will protect. The MAI protects all rights acquired by foreign investors, regardless of the inflow of capital into the host country. The term "investment" in the MAI broadly includes every kind of asset owned or controlled directly or indirectly by an investor; these include intangible assets, state authorizations or licences, claims to money, and all kinds of contractual rights. A proposed interpretative note in the MAI is expected to clarify that an "asset" must have "the characteristics of an investment, such as the commitment of capital or other resources, the expectation of gain or profit and the assumption of risk."

Furthermore, in a provision that extends well beyond the majority of existing bilateral and plurilateral investment regimes, the MAI prohibits "pre-establishment" discrimination. This would prohibit host governments from screening foreign investors prior to their establishing an investment presence.

The MAI will take a "top-down" approach to liberalizing investment rules. In other words, countries joining the MAI that wish to preserve their sovereign discretion to make distinctions between foreign and domestic investors must specify measures or sectors of their economy that they wish to shield from MAI disciplines, and formally lodge these with the OECD secretariat as country-specific exceptions. The exceptions will then probably be subject to "stand-still" and "roll-back" requirements that discourage Parties from adding to their exceptions. It is also probable that, through negotiations, political pressure will be applied to Parties to eliminate exceptions gradually over time.

Finally, the MAI's rules would be backed by compulsory state-tostate and investor-to-state dispute-settlement procedures. These allow both private investors and their home states to enforce the MAI through the binding judgements of an international arbitral tribunal.

Potential interaction between the CDM and the MAI

The MAI would support the CDM's general objective of promoting flows of capital from developed to developing countries. However, depending on how the details of the CDM rules are designed, there is some potential for conflict between the two regimes. Reference is made here to an initial analysis of this potential undertaken by the OECD Secretariat.

Broad scope of the MAI

The MAI would clearly have jurisdiction over any CDM project activity carried out in the territory of a developing country that chose to become a contracting party. Both a CDM project activity and any CER it might generate would fall within the broad scope of the MAI 's definition of an investment. A CER has the characteristics of an investment that is, the commitment of capital or other resources (i.e. technology transfer), the expectation of gain (i.e. the generation of profits and of CERs), and the assumption of risk (i.e. the risk that the project will not generate CERs). The CER may be a form of debt, such as a financial instrument, or a right conferred pursuant to law or contract, such as a government authorization or permit.

The MAI 's broad definition of investor would extend rights to all private entities or state-owned enterprises involved in a CDM transaction. It would not, however, include investments made by states themselves. States are not considered to be in need of any additional protection as investors, and would avail themselves of the diplomatic channels or the state-to-state dispute-settlement or non-compliance procedures under the Protocol to defend their rights.

Non-discrimination: MFN and NT

The MAI prohibits both de facto and de jure discrimination by the host state between foreign and domestic investors (the NT standard) and between two foreign investors from different states (the MFN standard). This means that host-country regulations that discriminate between these categories of investors either expressly, or in their effect, would be open to challenge under the MAI by either states or investors.

A potential for conflict may arise if a Party hosting a CDM project is encouraged or required by the Protocol to discriminate expressly between investors on the basis of the status of their home country in at least four ways, as outlined below.

1. Annex I versus non-Annex I Parties

Although not expressly prohibited by Article 12, it is unclear whether investors from non-Annex I Parties would be entitled to participate in CDM activities. Under some conceptions of "additionally, " project sponsors may be required to demonstrate Northto-South flows of financial resources before a project activity could be certified. In such a case, investors from non-Annex I Parties might be denied access either to eligible project activities or to CERs. It may be argued that a non-Annex I investor, without emissions-reduction commitments of its own, would have no incentive to invest in CDM projects. However, if CERs are designed as a tradeable commodity, it is entirely possible that an investor without commitments of its own would see the investment potential in buying and holding CERs to sell to the highest bidder should supplies become scarce.

2. Complying versus non-complying Parties

The Protocol Parties may wish to condition an investor's eligibility to participate in CDM activities on the basis of whether its home country is currently in compliance with its commitments. Article 6 of the Protocol (joint implementation; JI) sets a precedent by suspending a Party's right to add emission-reduction units (ERUs) generated by an JI project to its assigned amount if issues are raised with regard to either the investor or the host state's compliance.

It may be argued that a Party to the Protocol that has authorized the use of such sanctions in general would be unlikely to invoke (or even legally estopped from invoking) the MAI to challenge such a sanction when it is applied against one of its own investors. However, the MAI's investor-state dispute-settlement procedures may allow the investor (who may care less about the niceties of international legal obligations) to challenge a measure, even if its government feels otherwise.

3. Party versus non-Party

Although this is not explicit in Article 12, most conceptions of the CDM would probably not allow investors from host countries not Parties to the Protocol (or, at the very least, those not Party to the Convention) to participate in the generation and sale of CERs. This would be justified both for enforcement reasons (as a non-Party host country could not be expected to hold its investor to comply with CDM rules) and also to give all potential host countries an incentive to join the Protocol.

Indeed, the OECD Secretariat's own analysis of potential conflicts between the MAI and MEAs that used quotas and permits noted that:

If quotas or permits are earned by enterprises as a return on participation (investment) in a pollution reducing project in a developing country, the question would arise as to whether the ineligibility for such a quota or permit (return) of enterprises of countries not Party to the system constituted a discriminatory measure of the project host. If the eligibility requirement were established by an international regime, that might be interpreted for MAI purposes to be a measure of each Party to it.

The OECD Secretariat qualifies the risk by suggesting that barring investors from non-Parties to the Protocol from eligibility may not be necessary, as a CER would have no value in the legal system of the investor's home country. This analysis is, however, based on the assumption that CERs would not have an inherent value as an investment that could be sold to investors in home countries where they did have value.

4. Foreign versus domestic investor

Under some conceptions of the CDM, a host country or its own domestic investors would be eligible to invest in CDM project activities without the involvement of any foreign investor. Foreign capital would flow only at the point when the CERs were ready to be sold. In order to promote an endogenous, climate-friendly, technology in a particular sector, a host country might decide to bar foreign investors from CDM eligible project activities in the same sector, at least until the domestic producer was prepared to compete with foreign rivals. The MAI prohibition on pre-establishment discrimination would preclude such an approach that would discriminate against foreign investors.

Performance requirements

Article 12 provides that CDM project activities should assist developing countries in achieving sustainable development, and should promote real, measurable, and long-term benefits. By some analyses, such criteria would lead a host country to require a CDM project activity to shorten the chain of production by using locally produced goods or services, to build domestic capacity by employing local citizens, or to bring about the transfer of technology to a local firm. These employment and performance requirements, even if imposed equally on domestic and foreign investors, would potentially violate the MAI.

The MAI 's prohibition on performance requirements can be softened in two ways. First, the enumerated requirements may be employed in circumstances where they are conditioned on the "receipt or continued receipt of an advantage. " If CERs generated by project activities are seen as being within the control and largesse of the host state, then conditioning their transfer to an investor on the basis of performance requirements may be permissible. Secondly, the MAI text has a specific environmental exception applicable to the provision on performance requirements. Modelled on Article XX of the General Agreement on Tariffs and Trade (GATT) 1994, the performance-requirement exception would allow measures that might otherwise have violated the MAI if the host country can establish that they are "necessary for the conservation of living or non-living exhaustible natural resources. "

Expropriation and Compensation

Direct expropriation

The transaction at the core of the CDM (Article 12 (3)) is described so ambiguously as to leave unanswered a fundamental question: who has rights to the CER, or the expectation of a CER, at what stage in the CDM cycle? This issue is of great importance from the standpoint of investor protection, in that efforts by the host state to control or retain a CER for various reasons may be characterized as an "expropriation" of the investor's "investment" as broadly denned in the MAI.

If emissions reductions resulting from a CDM project activity were treated consistently with the regime established for Annex I party emissions, they could be characterized as "belonging" to the host government. Each Annex I Party has been "assigned" emissions allowances, and will ultimately be held responsible for emissions and credited for reductions that take place within its territory. Annex I Parties may, under some proposals for emissions trading, then devolve the right to these allowances to individual domestic emitters or investors. Were a similar construct applied to the emissions reductions achieved within the territory of a CDM non-Annex I host, the disposition of CERs would lie within the power of the host government. Host government regulations affecting the validity or value of a CER might not in these circumstances be considered an "expropriation."

Alternatively, Parties may determine that host countries should be entitled to retain a share of any CERs generated within their territory. Some have argued that a host should be able to collect a "resource rent" for maintaining the regulatory framework necessary for hosting the project activity. If standard rules are not agreed among the Parties on this issue, disputes might arise over the ad hoc expropriation of all or some of the CERs expected by an investor.

CDM rules may, indeed, increase the need for selective expropriation of CERs. As part of either a domestic or an international compliance regime, a host country acting of its own volition or on instruction from a Protocol body might suspend the validity of a

CER. As has been indicated, Article 6 of the Protocol (JI) sets a precedent by suspending a Party's right to add ERUs generated by an JI project to its assigned amount if issues are raised with regard to either the investor or the host states' compliance.

Indirect expropriation

The scope of the MAI's definition of expropriation would set a new global standard. Regulatory takings, or state measures such as taxation and licensing, which may affect foreign investments, do not traditionally amount to expropriation unless they are discriminatory or have the precise intent and effect of confiscation. The MAI, like the North American Free Trade Agreement (NAFTA) upon which it is modelled, expands the international standard for expropriation to cover "regulatory" takings. The MAI prohibits the taking of any state action or measure that has the equivalent effect of direct or indirect nationalization or "creeping" expropriation. There is standing available to an investor concerning an alleged breach of an MAI obligation which "causes loss or damage to the investor or it investment."

Whether the MAI would require compensation for the passage of regulations that reduce the potential for generating profits, or otherwise cause loss or damage to the investment, is a matter of current debate. The experience with NAFTA to date demonstrates that the current wording of the expropriation provision would support these claims. The liberalized MAI imposes broad obligations on states and new rights for investors. Together, this increases the possibility that any state regulation will directly or indirectly discriminate against one or more investors/investments. With broader grounds for discrimination, and a high standard of compensation, investors' rights to dispute-resolution mechanisms against states will undoubtedly influence domestic policy development under an MAI regime. This chapter now turns to two scenarios that test at a deeper level the potential relationship between the CDM and the MAI from the perspectives of a non-CDM investor and a CDM investor.

MAI and the non-CDM investor

The CDM, as with all environmental regulatory instruments, may be vulnerable to attack under the MAI if it provides the basis for any facially neutral regulation that has a disproportionate impact on a foreign investor. In this context, CDM-related rules aimed at strengthening a host government's regulatory framework would be as vulnerable to attack as any new environmental regulation that threatened the expectations of a foreign investor.

For example, every CDM project activity must achieve environmental additionally in order to be certified. This means that the project activity must bring about overall benefits that would not have occurred in the host country in the absence of the project. A counterfactual baseline, or reference case, must therefore be constructed (on either a multilateral or bilateral basis) to describe what the host country would have done in the absence of the project activity. The counterfactual baseline must be reliable and verifiable, in order to achieve the global reduction of GHG emissions. If CDM emissions reductions that are not additional are allowed to be certified and are offset against Annex B commitments, overall global emissions will increase against a business-as-usual baseline.

The COP/MOP will probably devise a common framework for determining baselines. The framework may be prescriptive (such as the existing framework for Global Environment Facility (GEF) projects, which requires that a project baseline must reflect at least a minimal standard of "environmental reasonableness") or it may give the Parties involved in the project activity more latitude when defining the baseline on a project-by-project basis. We assume the latter for this analysis.

Once a baseline is established between the host country and the Annex I investor, and a project activity has been certified, the host country may choose to adopt regulations that support the baseline so that the project activity will be verified and produce CERs. In order to prevent the reductions achieved from a CDM investment to be wiped out by emissions increases in the host country, CDM rules may require the host government to demonstrate within a broad systems boundary that dirty technologies are not being introduced elsewhere.

Analogies can be drawn from the NAFTA challenge of a US-based company, Ethyl Corp. (Ethyl), against the Canadian government for enacting legislation to ban the import and interprovincial transport of the gasoline (petrol) additive methylcyclopentadienyl manganese tricarbonyl (MMT) on the grounds that it is a dangerous toxin. MMT is added to gasoline to enhance octane and to reduce engine "knocking." Ethyl is the only North American producer of MMT, and a Canadian company directly benefited from the ban on MMT. Ethyl sued the Canadian government for approximately US$250 million, arguing that MMT is safe and that Canada's ban on the additive constitutes an illegitimate expropriation of Ethyl's assets namely, its Ontario plant, which conducted the final mixing of MMT. The Canadian government challenged the jurisdiction of the panel to hear the case, on the grounds that Ethyl followed improper procedure in bypassing their state government in initiating the claim. The panel ruled against them, finding that Ethyl had standing under NAFTA provisions, which are almost identical to those in the MAI. Shortly thereafter, the claim was settled for approximately US$13 million and an apology by the Canadian government.

The former Chairman of the MAI negotiations has recommended an addition to the draft MAI text of an "Interpretative Note" that would seek to prevent efforts by foreign investors to chill legislation that is facially neutral but that may have disproportionate impact. A footnote to the text would clarify that the MAI should not inhibit "normal non-discriminatory government regulatory activity." It also clarifies the presumption that it is within each Contracting Party's discretion to determine what measures "it considers appropriate" to protect health, safety, and the environment. However, the usefulness of the footnote is seriously undermined by the concluding phrase "provided such measures are consistent with this agreement," which restricts the Contracting Party's discretion to the bounds of the MAI obligations on NT, MFN, and expropriation. In other words, the paragraph does not provide an exception for environmental measures but, rather, what could, at best, be described as a clarification of burdens of proof. The burden of proof would clearly be on the complainant to establish that a facially neutral measure is inconsistent with the MAI.

Similarities between the MAI and GATT principles and provisions assist understanding of this issue. The Chairman's Note includes a proposal to remove the brackets that currently appear around the phrase "in like circumstances" in the Negotiating Text of the MFN and NT provisions and to add an interpretative footnote. This proposal is in direct response to news of the Ethyl dispute. This note is intended to allow a policy maker or an adjudicator of a dispute to consider "all relevant circumstances, including those related to a foreign investor and its investments, in deciding to which domestic or third country investors and investments they should appropriately be compared. "

In substance, the Chairman's approach introduces a concept similar to the "like product" concept under the GATT. When an investment-related measure is challenged as discriminatory, the term "like circumstances" invites a comparison between the domestic and foreign investors/investments, or between two or more foreign investors/investments, to determine whether the "the circumstances" related to the investors/investments justifies a distinction being made between them. Should the circumstances prove to be "like, " the discriminatory treatment would, presumably, violate the MAI.

The Interpretative Note anticipates two types of differential treatment that would be permissible with the introduction of the "like circumstances" language: (1) facially discriminatory measures that are justified by "legitimate policy reasons"; (2) measures that are facially neutral but that differ in their effect.

One can see emerging from the relationship between the phrase "like circumstances" and the Interpretative Note, a similar dynamic to that between the GATT's NT and MFN provisions, and the general exceptions in Article XX. Indeed, the Interpretative Note provides an example of the type of measure that, with the inclusion of the phrase "like circumstances, " would be considered as justifying an exception to the MAI (i.e. one "needed to secure compliance with domestic laws that are not inconsistent with national treatment and most favoured nation treatment. " In essence, this repeats the general exception in Article XX (d).

Thus the Interpretative Note, in effect, introduces a very broadly worded general exception to the MAI's NT and MFN requirements. An analysis of whether the "like circumstances" exception applies, would appear to require, in the context of a facially discriminatory measure, an assessment of whether the policy reasons behind the measure are "legitimate". In the context of a facially neutral measure, a complaining party or investor could not rely solely on the effect of the measure to demonstrate a violation of the MAI.

However, unlike Article XX, there is no direct guidance in the text (other than the Article XX (d) language), as to what types of policy reasons (such as conservation of natural resources) the negotiators consider "legitimate," or by what standard they should be judged legitimate (not arbitrary or unjustifiable). Given the "package" nature of the Chairman's Note, a strong argument could be made that the provisions on labour and the environment in the preamble provide guidance as to what is "legitimate," but this is by no means clear.

Left as a footnote to the text, the Interpretative Note takes on a questionable legal character. The Interpretative Notes to the GATT 1947 have played an influential role in the interpretation of that agreement, in part because they were explicitly included, through an Annex to Article XXXIV of the GATT, as "an integral part of" the Agreement.

Generally, the breadth and vagueness of the Interpretative Note could prove attractive to environment and labour groups, as they seem to provide considerable scope to the policy maker to determine individually what types of policies are "legitimate." Presumably, measures taken in accordance with MEAs would fall within what the Interpretative Note to the NT/MFN provision describes as "legitimate," but this has not been made clear. However, in a regime that has, as its primary objective, the promotion of investor protection and liberalization, the risks are high that, in the absence of clear guidance from either the content or the legal character of the Interpretative Note, the concept of "like circumstances" will be interpreted to prevent legitimate distinctions between investors/ investments. The inclusion of a positive right to protect the environment (subject to GATT-based disciplines prohibiting arbitrary, unjustifiable discrimination or disguised restrictions) is a far preferable option for allowing MAI contracting parties to pursue legitimate policy goals without being open to unrestricted challenges. Potential conflicts with MEAs, such as the CDM, could be addressed more directly through a specific provision on MEAs in a General Exception clause applicable to the entire MAI.

MAI and the CDM Investor

In another scenario, the CDM project activity may itself be expropriated by a host state. A host state may decide to nationalize a major industry or natural resource for purposes of social and economic development. For example, CDM project activities may include land-use change and forestry activities undertaken to reduce carbon emissions or increase carbon sequestration. Deforestation activities lead to combustion and decomposition of woody material and release carbon. Where land is purchased through a CDM project activity for the promotion of growth and regeneration in secondary forests and on pasture lands, and deforestation is prevented, the additional carbon that is sequestered may generate CERs.

A change of government and/or change in priorities or circumstance may cause a host state, after agreeing to participate in a carbon-sequestration project, to nationalize the land for other purposes. In addition to claiming expropriation of the land, an investor of a forestry CDM project will be likely to allege the expropriation of the CER certification. For the expropriation of the CER, the investor is likely to claim compensation for both the value of the land and the anticipated value of the offsets.

A host country could, therefore, face costs that include the provision of the same or similar land for a CDM project activity and compensation for present and future lost gains during the period of expropriation (or pecuniary compensation in lieu thereof). Additional benefits attached to the land itself as a CDM project activity (i.e. additional monetary investment; research for baseline; technology transfer) may also be claimed. The MAI requires compensation "equivalent to the fair market value of the expropriated investment immediately before the expropriation occurred. " In fact, the Ethyl case demonstrates how investors consistently argue (without exception under NAFTA) that an investment had additional value, both tangible and intangible, which goes uncompensated. Ethyl had claimed that the legislative debate in the Canadian Parliament itself constituted an expropriation of Ethyl's assets because public criticism of MMT damaged the company's reputation. US trade officials responded by arguing that the ability of investors to use legal threats to influence legislative debates is a healthy innovation that will prevent governments from passing laws that violate international agreements.

MAI Investor-State Dispute Resolution

The investor-state dispute-settlement procedures in the MAI would fill a significant gap in the Protocol's institutional structure. At present, it is anticipated that only Parties to the Protocol would have the power to invoke any of the Protocol's non-compliance or dispute settlement procedures. The MAI would be the first international agreement, open for accession to the global community, to give investors new rights and states additional obligations and to provide a mechanism for investors to enforce these rights through international arbitration. In extending these rights, the MAI may shift the balance of power between investor and host government in a way that was not anticipated when the climate negotiators designed the COM.

Each contracting Party to the MAI gives its unconditional consent to international arbitration in accordance with Article 4 upon signing the agreement. Any issue in dispute with respect to an alleged breach of an obligation (i.e. unlawful expropriation of a CER or CDM project activity) under the MAI that causes loss or damage to the investor or its investment, shall be decided in accordance with the MAI, interpreted and applied in accordance with applicable rules of international law. A Party to the MAI would further have to agree to submit any other investment dispute concerning any obligation which the host state has assumed pursuant to the agreement to enter into a CDM project activity or transfer a CER (which can be considered an investment authorization) to arbitration. In this case, the rules of law agreed to by the Parties under the agreement would prevail. Where the agreement is silent, the law of the Contracting Party and applicable rules of international law prevail. The investor may choose to submit any dispute that cannot be settled through negotiation or consultation to a number of specified forums. Only the investor has a right to withdraw a dispute once it has been initiated. In the absence of an investor's right to bring a claim directly, domestic industries or multinationals would put pressure on their home governments to bring actions to protect their commercial interests. Specious claims that may not be in the bilateral or global interest are usually filtered out by the home government. In the absence of this government filter, and on the basis of the experience of NAFTA, the MAI will be likely to increase international arbitration.

Summary

The agreement on Article 12 resolved a number of critical aspects of how the CDM will manage project-based JI between Annex I and non-Annex I Parties to the Protocol. However, many gaps remain to be filled, and the negotiating dynamic for the next stage of the development of the CDM remains fundamentally unchanged. This dynamic can now be characterized as pitting a market-based approach against an "interventionist approach" based on traditional public-sector development assistance. Both approaches stress the need for a system capable of generating credible CERs but differ on the best means of achieving this. The extent to which the CDM might fall foul of the MAI will depend on the level of state intervention that Parties feel will be necessary to achieve the CDM 's policy goals.

A market-based approach relies upon healthy competition in a transparent market-place to provide the most efficient and effective means of encouraging hosts and investors to design credible CDM project activities. Once the intergovernmental process has set the rules on the types of project activities that will be eligible for certification, the private sector which holds the capital and technology necessary to the CDM's success would be entrusted with designing projects and would be entitled to hold and transfer CERs.

Interventionists are more sceptical of the private sector's ability to fulfill the CDM's stated purpose of assisting non-Annex I Parties to "achieve sustainable development. " Such an approach emphasizes the need for the active involvement of public-sector institutions, including home and host governments and international development institutions, in promoting the design of projects driven by broad based policy concerns rather than market disciplines.

The debate is further complicated by the tension between those that wish to see the CDM up-and-running quickly and with as low transaction costs as possible, and those that remain cautious and are willing to increase costs in exchange for greater accountability. Parties at both ends of this spectrum place the CDM at risk, either by undermining its credibility or by weighing it down with an over burdensome bureaucracy.

Theoretically, conflicts between the Protocol and the MAI could be handled through a system of country reservations, whereby any Party to the Protocol that chose to accede to the MAI would "opt out" any relevant measures. If applied liberally, this could allow the Contracting Parties to the MAI to exempt from the Agreement's NT and MFN rules legitimate existing measures that facially discriminate between domestic and foreign investors, in order to achieve a variety of policy goals, including environmental protection. For example, the Parties to NAFTA, under a similar legal arrangement, have exempted their fisheries and other environmentally related sectors from that Agreement's NT obligations.

However, experience under other liberalization regimes suggests that it is difficult to draft exemptions or waivers precisely enough to avoid challenges based on such broad principles as NT and MFN. The continuing challenge to the MAI negotiators and to the designers of the CDM will be to coordinate efforts in both regimes to ensure that the potential for conflict is diminished.

Even if the MAI is never adopted, capital-exporting countries are likely to continue to exert pressure for the acceptance of high standards of investor protection in other forums, either bilaterally or multilaterally. Coming to agreement on crucial issues, such as the rules on the ownership or sharing of entitlements to CERs generated by CDM project activities, and the dispute-settlement procedures available to private entities, are likely to prove important to the CDM's stability and success.

5

Climate Change Convention

Introduction

The procedures is evidence of a growing awareness that traditional rules of international law concerned with material breach of treaty obligations and with state responsibility are inappropriate to address problems of environmental treaty implementation. It is the purpose of this chapter to set forth some of the issues that will need to be addressed in establishing an NCP under the Climate Change Convention, in particular in connection with the application of such procedures to the mechanisms for achieving the substantive commitments undertaken by developed States under the Kyoto Protocol. Before the Climate Change Convention and the Protocol are considered, the concept of NCP and the development of such procedures in contemporary international environmental law is traced.

What are "non-compliance procedures"Essentially, NCPs embody procedures established under multilateral treaties to meet the following objectives:

1. To provide positive encouragement to Contracting Parties to comply with their treaty obligations.
2. To provide a multilateral forum for dispute resolution/ avoidance.

3. In the event of non-compliance, to provide a "softer, " less legalistic mechanism than offered by traditional dispute-settlement procedures under international law. NCPs generally apply a less rigid test by which compliance is measured and the first object is to obtain a return to full compliance by the "defaulting" state, rather than impose a sanction for non-compliance or award compensation to an injured party.
4. To provide for internal resolution of disputes, without recourse to external adjudicators or institutions.
5. To facilitate access, since NCPs may typically be invoked by, inter alia, one Party, and are therefore not dependent upon common agreement; thus, unlike traditional dispute-settlement mechanisms, NCPs need not be consent-based.

From these objectives, it may be seen that NCPs are concerned both with facilitating compliance by Contracting Parties with their Convention obligations, and with providing a "softer system" for addressing non-compliance by a Contracting Party than is currently afforded by traditional dispute-settlement procedures under international law. In the latter context, NCPs may be viewed as an international form of alternative dispute settlement (ADR), long employed in municipal legal systems to resolve conflicts arising in longer-term legal relationships. The use of NCPs at the international level is thus a recognition of the relational character of certain international treaties and the need to ensure both continuing participation and the fulfillment of generally non-reciprocal treaty obligations designed to ensure the achievement of common goals. Accordingly, NCPs are less confrontational and generally less legalistic and are designed to assist the defaulting State in returning to compliance, not necessarily to incriminate for non-compliance. This is reflected in a trend away from provisions relying solely on "dispute settlement, " a phenomenon observable not just in the environmental sphere, towards the addition of provisions addressing "assessment," "monitoring, " "verification, " "verification of compliance, " implementation, " and "considering progress made in the implementation" of the Convention. Many of the preceived advantages of an NCP are relected in the rationale fo the multilateral consultative process set forth in Article 13 of the Climate Change Convention.

The dual nature of NCPs is also a reflection of the various reasons for non-compliance by states with their international obligations, which may range from the free rider exploiting the economic advantage derived from non-compliance to an inability to meet treaty obligations because of their high cost or a lack of capacity, e.g. lack of relevant technology or expertise. An NCP needs both to reduce/eliminate the economic benefit to be derived from non-compliance and to facilitate compliance where obstacles relating to lack of capacity, particularly for developing states, are identifiable.

The goal of NCPs is therefore distinguishable from traditional rules on state responsibility, since the object is to ensure a return to full compliance with treaty obligations rather than to require compensation by the defaulting state for the harm caused to another state or states for breach of an international obligation. The latter point highlights a further negative feature of traditional dispute-settlement mechanisms in the environmental context namely, that environmental damage has generally already occurred. Moreover, in the climate-change context, the effects of non-compliance may be "Subtle and commutance," with the full effects manifest only over a considerable time. In common with other transfrontier environmental problems, it may also be difficult to attribute particular harm to the actions or omissions of one state or several states. This enhances the need for a NCP that places the emphasis on prevention rather than on remedying harm caused.

Although they are not confined to the environmental context, it is none the less clear that NCPs have a particularly important role to play in that context because of the distinctive character of environmental treaty obligations. A number of factors may be identified: the pace, magnitude, and irreversibility of environmental problems; the essentially non-reciprocal nature of the treaty obligations which renders enforcement inter partes ineffective; the failure to operationalize traditional rules on state liability and responsibility; the ability to measure compliance against quantifiable targets; and the necessity for national implementation to render international environmental obligations effective. Perhaps the best-known existing NCP is that established under the 1987 Montreal Protocol to the 1985 Vienna Convention on the Protection of the Ozone Layer ("Ozone Convention"). This was the first multilateral environmental treaty to go beyond reporting and dispute-settlement provisions in providing for a non-compliance procedure. While Article 11 of the Convention provides for traditional dispute settlement, Article 8 of the Protocol envisages an NCP, which has subsequently been developed by meetings of the Parties. The procedure was finalized at the Fourth Meeting of the Parties (MOP) in London in 1992. A pivotal role is played by a ten-member Implementation Committee. Early issues included data reporting to establish important baselines for assessing compliance with quantitative and temporal obligations (targets and timetables).

Monitoring and compliance are essential bedfellows. There is little point in adopting a compliance mechanism without the capability to gather information relevant to a determination whether treaty obligations are being met. This is part of an integrated process that has been termed "implementation control." This can take place in incremental stages, i.e. monitoring and information gathering preceding the development of non-compliance mechanism(s). A good example of this process is the 1979 Geneva Convention on Long-Range Transboundary Air Pollution (LRTAP), which has seen data collected under the EMEP Programme form the basis for an assessment of the first sulphur-emissions protocol in 1985 and led to agreement on a tightened emission-reduction schedule in the second sulphur protocol in 1994 with a non-compliance mechanism. Of course, the obligation to report is itself a concrete obligation, non-compliance with which may be subject to an NCP.

If such information gathering and reporting is viewed as part of a broader compliance-information system, the question remains of what response is made, if any, to identified problems of non-implementation by Parties. In the absence of a non-compliance mechanism, non implementation can be addressed only (in formal legal terms at any rate) as a dispute between contracting Parties, dealt with if at all under traditional dispute-settlement mechanisms. This is the present position under the 1992 Convention on Biological Diversity it is the Parties themselves who inspect reports, with any alleged nonimplementation a matter for dispute settlement under Article 27 of the Convention. Other recent environmental treaties go further these include the 1987 Montreal Protocol; the 1991 Protocol Concerning the Control of Emissions of Volatile Organic Compounds or their Transboundary Fluxes (VOC) and 1994 Sulphur Protocols to the 1979 LRTAP Treaty; the 1992 Convention for the Protection of the Marine Environment of the North-East Atlantic; and, of course, the 1997 Kyoto Protocol, all of which envisage a separate mechanism for compliance being established. The Parties to the 1989 Basel Convention on the Transboundary Movement of Hazardous Wastes and their Disposal are also considering an NCP. Of these examples, only the NCP under the Montreal Protocol is fully operationalized.

Without concrete obligations, and the monitoring of implementation/compliance with those obligations, it is not practicable to speak of non-compliance mechanisms. This is seen clearly in the climate-change context, which has proceeded from general commitments without specific timetables and targets to agreement upon such specific commitments with attention turned to the implementation thereof and to compliance control. The 1987 Montreal Protocol, the 1994 Second Sulphur Protocol, and the 1997 Kyoto Protocol suggest a new cycle of strengthening compliance systems gradually, and in step with the strengthening of commitments.

The FCCC

The traditional dispute-settlement approach is present in Article 14, which adopts an approach that conceptualizes disputes as arising between two or more Contracting Parties in connection with the interpretation or application of the Convention. If ever activated, the traditional hierarchy of peaceful dispute-settlement mechanisms would apply, ranging from negotiation and third-party mediation or good offices, through to arbitration or submission of the dispute to the International Court of Justice. Under Article 14, recourse to negotiation or other means of peaceful settlement is obligatory, with either Party able to request creation of a conciliation commission in the event that negotiation is unsuccessful; however, the awards of the commission are recommendatory only. This bilateral dispute-settlement route is considered to be complementary to the Article 13 process.

Article 13, on the other hand, is a good example of the trend away from total reliance on traditional dispute-settlement methods in recent multilateral environmental agreements noted above. It establishes a multilateral consultative process (MCP) for resolution of questions concerning the implementation of the Convention. Little detail is contained in Article 13; thus, the first meeting of the COP established an Ad Hoc Group on Article 13 to operationalize the MCP. The sixth and final session of this Group was held in Bonn in June 1998, where its work was completed in anticipation of COP 4. It has adopted the framework for an MCP which must now be considered at COP 4, including the resolution of the matters left unresolved in the Committee. This will clearly encompass the commitments contained in Articles 4–6 and 12 of the Convention. Of particular note is the objective of the process, set forth in Paragraph 2, and the mandate of the proposed standing Multilateral Consultative Committee, set forth primarily in Paragraph 6:

2. The objective of the process is to resolve questions regarding the implementation of the Convention, by:
   a. Providing advice on assistance to Parties to overcome difficulties encountered in their implementation of the Convention;
   b. Promoting understanding of the Convention;
   c. Preventing disputes from arising.
6. The Committee shall, upon a request received in accordance with paragraph 5, consider questions regarding the implementation of the Convention in consultation with the Party or Parties concerned and, in light of the nature of the question, provide the appropriate assistance in relation to difficulties encountered in the course of implementation, by:
   a. Clarifying and resolving the questions;
   b. Providing advice and recommendations on the procurement of technical and financial resources for the resolution of these difficulties;
   c. Providing advice on the compilation and communication of information.

In earlier sessions the Group has emphasized the advisory rather than supervisory nature of the MCP, further distancing the process from a more rigorous form of NCP. This is clearly reflected in the language of Paragraph 6 which is strongly facilitative: "appropriate assistance, " "clarifying and resolving, " "providing advice and recommendations, " and "providing advice. " The advisory nature of the MCP is further enforced by the link between Paragraphs 6 and 12. Under Paragraph 12, the Committee may make "[recommendations regarding cooperation between the Party or Parties concerned and the other Parties to further the objectives of the Convention", and "measures that the Committee deems suitable to be taken by the Party or Parties concerned for the effective implementation of the Convention." Whilst specific outcome is not expressed in mandatory terms ("may"), Paragraph 12 is mandatory in the requirement that the outcome of the Committee's deliberations must be consistent with its mandate in Paragraph 6.

The MCP is without prejudice to the dispute settlement provisions of Article 14, the latter applying, mutatis mutandis, to the Protocol. There is no internal "exhaustion of local remedies rule" in operation. However, there is some doubt whether Article 14 will ever be invoked in a traditional dispute settlement; indeed, one of the reasons for including Article 13 was the perception that traditional dispute settlement would have a very limited role to play under the Convention where the likely nature of disputes would not be amenable to such procedure. None the less, the relationship between the two requires some clarification. Some states favoured the automatic suspension of Article 13 when Article 14 is invoked, but in the event this is not the approach of Paragraph 4 of the terms of reference of the MCP, which stresses that the process is both separate from and without prejudice to the provisions of Article 14 of the FCCC. Whilst this addresses the procedural hierarchy of the Article 13 and 14 processes, it may not wholly cover the substantive interaction of disputes arising under Article 13 and Article 14 procedures. For example, the "without prejudice" argument might be difficult to maintain where non-conforming behaviour has been condoned under an Article 13 mechanism but an Article 14 dispute none the less arises in connection with that act or omission. Though it is beyond the scope of this chapter fully to explore the issue, condonation of behaviour in technical breach of the treaty could give rise to acquiescence and estoppel arguments being maintained against the complaining state.

NCP under the 1997 Kyoto Protocol

The application of the Article 13 MCP to the Protocol is an issue left undetermined by the Protocol itself. Article 16 of the latter provides that the COP serving as a meeting of the Parties to the Protocol shall "as soon as practicable" consider such application, with or without modification. If the Article 13 MCP were so extended, the Protocol expressly provides that such procedure would operate without prejudice to the NCP under the Protocol (which in turn is without prejudice to the dispute-settlement provisions of Article 14 FCCC). What is clear is the determination to distinguish a specific NCP under the Protocol from both the dispute settlement provisions of Article 14 of the Convention/Protocol and the MCP of Article 13 (if extended to the Protocol).

It is Article 18 of the Protocol that expressly refers to noncompliance in the following terms:

The Conference of the Parties serving as the meeting of the Parties to this Protocol shall, at its first session, approve appropriate and effective procedures and mechanisms to determine and to address cases of non-compliance with the provisions of this Protocol, including through the development of an indicative list of consequences, taking into account the cause, type, degree and frequency of non-compliance. Any procedures and mechanisms under the Article entailing binding consequences shall be adopted by means of an amendment to this Protocol.

Owing to the politically sensitive nature of NCPs in particular, binding consequences flowing from a determination of noncompliance it is not surprising that decision on any such characteristics will require the stringent treaty-amendment procedures of the Protocol to be followed. Establishing this significant procedural hurdle to the adoption of binding consequences for noncompliance is, in part, a reaction against the dynamic development of the Montreal Protocol NCP unfettered by such further requirement of treaty amendment but subject rather to the decision-making rules of the COP. There is a clear reluctance to provide a "blank cheque" for binding non-compliance consequences to the COP, even as the supreme body of the Convention.

The design of a future NCP under the Protocol will entail both institutional and functional aspects: what is the procedure designed to achieve, and which organ(s) will be responsible for it? A special body will need to be established, probably a standing committee of legal, economic, and technical experts (or generalists with access to a roster of experts). Moreover, the relationship between this body and the existing Convention bodies will require careful definition. Experience has shown that the development of an NCP can take some time. How, then, will the NCP be operationalized pending the entry into force of the Protocol? This is particularly problematic, given that the key features of the flexibility mechanisms under the Protocol have also yet to be determined.

In fact, these matters have already been raised in the Subsidiary Body for Scientific and Technological Advice (SBSTA) and the Subsidiary Body for Implementation (SBI) which have met since the adoption of the Kyoto Protocol (Bonn, June 1998). Included on the agenda of each body was consideration of suggested elements for a work programme to operationalize the mechanisms under the Kyoto Protocol in particular, joint implementation (JI), the Clean Development Mechanism (CDM), and emissions trading. Compliance is identified as one of the outstanding issues under each of these mechanisms, addressed in turn below.

Mechanisms under the Kyoto Protocol

Under Article 6 of the Protocol, Annex I countries may acquire "emission-reduction permits" resulting from projects with other Annex I Parties aimed at reducing emissions or enhancing natural carbon sinks. Such JI is, however, expressly stated to be supplemental to domestic action to curb greenhouse-gas (GHG) emissions; thus, it is not possible for Annex I Parties wholly to meet their Article 3 reduction commitments via this route. Nor is it possible to buy up emission-reduction units (ERUs) to meet Article 3 commitments where a question of implementation has arisen in the course of the expert review of annual inventories. Key elements of JI remain to be addressed by the first meeting of the COP serving as a meeting of the Parties to the Protocol, most notably guidelines on implementation, including verification and reporting procedures. The extent to which ERUs may be utilized as "supplemental to domestic actions" also requires clarification, given concerns that the flexibility of Article 6 may be used to avoid meeting Article 3 commitments through domestic measures. Not least, this raises competition concerns, with the "burden on business" perceived to be less where an Annex I Party relies extensively on ERUs in meeting Article 3 commitments.

Where JI is pursued, Article 4 provides that a failure to achieve joint emission-reduction targets (ERTs) does not absolve Parties from the obligation to meet their own ERTs which are obliged to be set forth in the agreement. This simplifies the application of an NCP to the JI process where there has been a failure to achieve targets, and provides additional incentive for reaching the targets set forth in the JI agreement. Whilst implementation may be joint, responsibility for non-compliance with targets is still that of the individual State.

The verification and reporting criteria which are to be established at the Meeting of the Parties to the Protocol (MOP) 1 (or as soon as practicable thereafter) could give rise to non-compliance thresholds, as could the extent to which JI is supplemental to domestic implementation measures. The additionality requirement of Article 6 (1) (b) provides a further benchmark for the application of an NCP.

Clean Development Mechanism

Article 12 of the Protocol establishes the CDM, the purpose of which is "to assist Parties not included in Annex I in achieving sustainable development and in contributing to the ultimate objective of the Convention, and to assist Parties included in Annex I in achieving compliance with their quantified emission limitation and reduction commitments under Article 3. " It will be recalled that, notwithstanding strong support from some Annex I Parties for a New Zealand proposal to make wealthier developing countries also subject to specific commitments, the Protocol introduces no new commitments for developing states. What Article 12 accomplishes is to provide a mechanism whereby JI between Annex I and non-Annex I Parties may take place, with Annex I Parties implementing quantifiable emission reductions under Article 3 of the Protocol while non- Annex I Parties implement the obligations set forth in the Convention. Private and/or public entities may participate in projects which are subject to certification and verification in order for emission reductions to be "claimed". The mechanism is to be supervised by an executive board. The executive board and "operating entities" under Article 6 could perform a compliance function in respect of the CDM, in which case the issue of whether multiple compliance mechanisms will evolve under the Protocol, perhaps linked with specific flexibility mechanisms, will need to be addressed by the

COP/MOP. As under Article 6, the establishment of auditing and verification criteria will also give rise to the need to establish non-compliance parameters.

As with JI between Annex I Parties, concerns that Article 3 commitments would be met wholly through the CDM are addressed in Article 12 (3) (b), which explicitly provides that certified emission reductions from such project activities may contribute to compliance with part of their Article 3 commitments (as determined by the COP). A key concern for all three of the Kyoto mechanisms is to establish an appropriate level of reliance on these mechanisms, jointly and severally, in addition to domestic implementation. The setting of a precise level would constitute a yardstick against which compliance with the supplementarity requirement could be measured. Finally, as with JI, emission reductions deriving from CDM project activities must demonstrate that such reductions are additional to any that would occur in the absence of the certified project activity [Article 12 (5) (c)}, thus providing a further benchmark for the application of an NCP.

Emissions trading

Article 3 of the Protocol envisages an emissions-trading system that will establish a market amongst Annex I Parties in emission credits. Thus, for example, if an Annex B Party discovered that it risked exceeding its quota, it would have the option of acquiring some or all of the unused quota of another Annex B Party, thus increasing its total allowable emissions under the Protocol. However, it is left to the COP to define "the relevant principles, modalities, rules and guidelines, in particular for verification, reporting and accountability for emissions trading. " Nor is any time-scale for operationalizing emissions trading stated in the Protocol, though it is certainly expected to be operational during the first commitment period (2008- 2012). Any trading is expressed to be supplemental to domestic actions to meet reduction commitments; limiting the operation of the trading to developed states further meets the concern expressed by developing states that such states would meet their quotas without implementing necessary domestic measures to reduce emissions, simply through purchasing quota.

Summary

A great deal has been left for determination by the first meeting of the COP serving as MOP I, including the NCPs and mechanisms referred to in Article 18. Moreover, as has already been observed, no binding consequences arising out of noncompliance may be imposed under the Protocol without further amendment of that instrument in accordance with the procedures set forth in Article 20 thereof. It is a tall order to establish the NCP at the first MOP, particularly given that so much of the detail of the Kyoto mechanisms remains to be fleshed out. In practice, it is SBSTA and the SBI that have been charged with the preparatory work to allow the COP/MOP to discharge its functions, including an Article 18 NCP. Further guidance from the Parties on the modalities and schedule for operationalizing the flexibility mechanisms will come from COP 4.

There are a number of factors which should be considered in establishing an NCP under the Kyoto Protocol. First is the list of general guiding principles that any NCP agreed should be simple, facilitative, cooperative, non-judicial, non-confrontational, timely, and transparent. In fact, these are many of the characteristics of the Article 13 MCP and are broadly consistent with the essential characteristics of an NCP identified by the Working Group charged with establishing the Montreal Protocol NCP. As for the functions of an NCP, these could include (a) data collection (e.g. national reports), (b) review of data, (c) investigation (e.g. further information requested, site visit), (d) recommendation(s) (to indicate the end of the review process, with full or partial compliance indicated), and (e) further measures (facilitative or sanctions). A new body (or bodies) would have to be created, along the lines of the Committee under the Article 13 MCP or the Committee on Implementation under the Montreal Protocol. Membership issues (e.g. size, duration of tenure, geographic distribution, and personal or government representation) could prove contentious, as has been the case for the composition of the Article 13 MCP Committee. For example, should the body reflect the traditional UN approach of equitable geographic distribution, or should other principles govern (such as weighting in favour of Parties undertaking specific timetables and commitments under the Protocol)? Who should take actual decisions, particularly in respect of binding consequences of non-compliance? Should this be the COP (the procedure followed by the Montreal Protocol and under the Second Sulphur Protocol)? One benefit of such division of function between an NCP body and the COP is that this encourages the NCP to be less confrontational and will facilitate cooperation between the NCP body and the "defaulting" state.

In terms of the measures taken by a NCP body in the event of non-compliance, studies of the most fully operational NCP the Montreal Protocol NCP have shown that a combination of facilitation and sanction (assistance and coercion) is the most effective in ensuring compliance with treaty obligations. Facilitative measures might include interpretation of ambiguous provisions and financial assistance to meet obligations arising under the Protocol. However, since the obligations undertaken are by Annex I Parties, lack of financial and technical resources is unlikely to be a convincing explanation for non-compliance.

A typical coercive mechanism is to reduce or withhold the benefits of treaty participation, including access to Global Environment Facility (GEF) funding or to the proposed Multilateral Carbon Fund, if established, but a sanction of little relevance to non-complying Annex I Parties. This is part of the application of a sanction of suspending rights and privileges enjoyed under the Protocol, which could range from limiting permitted reliance on ERUs under JI and the CDM to reducing or denying access to an international emissions-trading system. There are few examples of multilateral trade sanctions being applied for non-compliance, a rare exception being the NCP agreed by the International Commission for the Conservation of Atlantic Tuna (ICCAT) in 1996. The compatibility of any measures adopted with other treaty obligations, including General Agreement on Tariffs and Trade (GATT)/World Trade Organization (WTO) obligations, will require further, careful analysis. Some fisheries arrangements provide for a reduction in total allowable catch for a subsequent fishing period, where the total allowable catch for the existing management period has been exceeded. An analogous approach under the Protocol would be increasing reduction commitments in a second commitment period; however, without the establishment of such a period, such a mechanism would be inoperative. Since the MOP is to institute consideration of Annex B at least 7 years before 2012 i.e. in 2005 this might be the stage at which to link consideration of future commitments with analysis of "significant progress"; however, any change in Annex B requires the amendment procedures of the Protocol to be complied with and the written consent of the Party concerned.

Article 18 specifically mandates the development of an indicative list of consequences of non-compliance with the provisions of the Protocol, "taking into account the cause, type, degree and frequency of non-compliance. " It has been suggested that an NCP could be simplified and shortened through the prior classification of noncompliance consequences automatic or discretionary depending on the type, degree, and frequency of non-compliance. This is closer to a coercive, sanction-based regime and is probably most suited to breach of quantifiable, non-derogable, and fixed obligations. As already indicated, however, if binding consequences are stipulated, amendment of the Protocol is required.

In addition to considering the prior classification of non-compliance responses, there is the additional possibility of a listing of presumptively non-complying acts and/or omissions. Indeed, since Article 18 links the development of an indicative list of consequences with, inter alia, the type of non-compliance, it is difficult to see how some classification of non-complying acts may be avoided if Article 18 is to be implemented fully. There are two, potentially compatible, approaches to setting non-compliance triggers. The first, a "bottom up" approach, might arise from monitoring and reporting requirement: for example, Article 3 (3) requires States to submit data regarding GHG sources and sinks, which are reviewed by expert review teams pursuant to Articles 7 and 8; such analysis of national reports under Articles 7 and 8 could be linked with the Article 18 NCP in a "bottom-up" approach to NCP. A "top-down" approach is the prior classification of certain acts and omissions as prima facie instances of non-compliance. This might include various temporal triggers for assessing compliance with the provisions of the Protocol, viz. 2005 [has demonstrable progress in implementing commitments been exhibited as required under Article 3 (2)?}, 2007 [has a national system for reductions been implemented as mandated by Article 5 (1)?}, and 2008–2012 [have emission reductions been met within the commitment period as stipulated by Article 3 (1)?}. An

6

Greenhouse-Gas Emissions

Introduction

The United Nations Framework Convention on Climate Change (UNFCCC) at the Earth Summit in June 1992, which committed Annex I countries only to "aim" to stabilize emissions of carbon dioxide (CO_2) and other GHGs at their 1990 levels by 2000, the so-called Kyoto Protocol sets legally binding emissions targets and timetables for these countries. Together, Annex I countries must reduce their emissions of six GHGs by at least 5 per cent below 1990 levels over the commitment period 2008– 2012, with the European Union (EU), United States, and Japan required to reduce their emissions of such gases by 8, 7, and 6 per cent, respectively. Although proposals had been made for differentiation of allowed emissions on the basis of indicators such as population, gross domestic product (GDP), or carbon intensity of the economy, the differentiated targets agreed upon at Kyoto were purely political. The Protocol will become effective once it is ratified by at least 55 Parties whose CO_2 emissions represent at least 55 per cent of the total from Annex I Parties in the year 1990. Pushed by the United States, the Kyoto Protocol also accepts the concept of emissions trading in principle, under which one Annex I country will be allowed to purchase the rights to emit GHGs from other Annex I countries that are able to cut GHG emissions below their assigned amounts (i.e. their targets); however, the Kyoto Protocol leaves the design of the market and its rule entirely to subsequent conferences. Structured effectively, the market-based emissions-trading approach, pioneered in the US SO_2 Allowance Trading System, can provide an economic incentive to cut GHG emissions while also allowing flexibility for taking cost-effective actions. It is generally acknowledged that the inclusion of emissions trading in the Protocol is in line with the underlying principles in Article 3.3 of the UNFCCC, which states "policies and measures to deal with climate change should be cost-effective so as to ensure global benefits at the lowest possible cost," and reflects an important decision to address climate-change issues through flexible market mechanisms.

As the successor to the General Agreement on Tariffs and Trade (GATT), the World Trade Organization (WTO) was created in 1994 upon the completion of the Uruguay Round of multilateral trade negotiations. Its Committee on Trade and Environment (CTE) has been established to coordinate the policies in the field of trade and environment. The Committee's work programme includes a review of "the relationship between the provisions of the multilateral trading system and trade measures for environmental purposes, including those pursuant to multilateral environmental agreements". Although emissions trading has been identified for future discussion in the Committee, it has not thus far been examined. It remains unknown whether WTO provisions would cover an emissions-trading scheme, in part because no interpretation exists on whether a legal definition of emissions trading would be interpreted as either trade in a good or trade in a service. No doubt, the inclusion of emissions trading in the Kyoto Protocol will catalyse the international consciousness for the potential of emissions trading. Clearly, this will provide a stimulus to addressing the market-based instrument in the Committee.

This chapter examines the relationship between GHG-emissions trading and the world trading system. The next main section explains why emissions trading is considered to be the most promising way to control GHG emissions, and the three subsequent main sections discuss, respectively, the basic requirements for setting up a successful emissions-trading scheme, some trade-related aspects of emissions trading, and joint implementation (JI) with developing countries.

Why Emissions Trading?

GHGs are uniformly mixed pollutants, i.e. one ton of a GHG emitted anywhere on earth has the same effect as one ton emitted somewhere else on earth. Translated into the language of abatement strategies, this means that it does not matter where reductions in GHG emissions take place; what matters is whether we are able to reduce the emissions effectively on a global scale. Given the fact that the costs of abating GHG emissions differ significantly among countries and that, unlike sulphur dioxide (802) emissions, there are no local "hot spots" for GHG emissions, GHG-emissions trading seems to enjoy an even better prospect for trading than SO_2. This large potential of efficiency gains, backed up with the widely regarded successful C_2 Allowance Trading System in the United States, conveys the message that emissions trading is a very attractive abatement option. How, then, does emissions trading compare with carbon taxes, another economic instrument that is widely believed to be able to achieve the same emissions target at lower costs than the conventional command and control regulations?

In economic theory, the two instruments can achieve identical results given both perfectly competitive markets and certainty. In practice, however, there could be major differences between these two instruments. The most valid arguments in favour of tradeable permits rather than taxes so far are as follows. First, tradeable GHG-emissions permits, unlike carbon taxes, are a form of rationing and the great advantage is that, in this way, one can be sure of achieving the target agreed. This feature seems to be appealing more than ever because Annex I parties to the Kyoto Protocol are obligated to comply with their legally binding emissions targets. This also makes the "ecological transparency" argument against emissions trading no longer valid. By contrast, the actual achievements in reductions of CO_2 emissions by a proposed carbon tax remain uncertain because of imperfect knowledge of the price elasticity of demand and supply for fossil fuels, especially for the large price increases caused by carbon taxes for major emissions cutbacks. This implies that setting the initial tax will be a hit-and-miss affair, and could thus induce hostile reactions from countries, industries, and consumers although it is not clear how serious an objection this is. Moreover, in the context of global warming, the delays in adjusting the insufficient carbon tax to the desired level will imply more CO_2 emissions emitted into the atmosphere than would otherwise have occurred, thus leading to additional committed warming.

Another complication of the carbon tax is the initial difference in energy prices. As a consequence of existing distortions by price regulations, taxation, national monopolies, barriers to trade, and so on, there are initially great differences in energy prices, both between fuels and across countries. If CO_2 emissions are then to be reduced by similar amounts in two countries, ceteris paribus, lower taxes are required for the country with low prices before the tax imposition than for the country with the higher pretax prices. Thus, an eventual cost-efficient regime of international carbon tax would presumably need to remove existing distortions in international energy markets otherwise, countries with the lower pre-tax prices would enjoy free-rider benefits.

Third, and most importantly, emissions trading offers a built-in feature of resource transfers by emission sources to developing countries. Such transfers are crucial to getting developing countries engaged in controlling GHG emissions. Of course, it is not impossible to include transfers in an international scheme of carbon taxes, but the trouble with it is that we need an international agency to collect carbon taxes. Given the fact that the United Nations still have difficulty in collecting their membership fees and that no other institution is of higher international jurisdiction than the United Nations, this will leave some doubt about its capacity to collect international carbon taxes. Even if such an agency manages to obtain the proceeds and uses them as transfers, there are still serious doubts as to whether it can efficiently manage such transfers.

Fourth, emissions trading is more attractive to firms than are carbon taxes, because the latter scheme extracts revenues from firms without offering any compensation, not to mention the political difficulties of introducing such taxes in countries such as the United States. So, even if a firm has to buy permits now to cover all of its emissions, it still can acquire the value of those additional permits by selling them in the future if its actual emissions are lower than the allowed limit. This in turn creates an incentive for firms to comply with their "caps".

Basic Requirements for Setting up Emissions Trading

Even if international emissions trading is considered to be the most promising way to control GHG emissions, then what are the basic requirements for setting up a successful scheme?

First, there should be legally binding national emissions targets and timetables for reducing GHG emissions for countries that would wish to participate in an international emissions-trading scheme. Those countries should be committed to the binding obligations.

Second, there should exist a reliable national registration of individual emissions sources that will participate in an emissions-trading scheme. Without such an inventory of sources and their present emission levels, it would be impossible to design schemes for permit allocation by way of grand fathering permits. Moreover, since countries (not sources) sign the Kyoto Protocol and it is the responsibility of the governments to ensure that their countries are in compliance with the national emissions limits, inter source trading would have to be accounted for at the national level. This also underlines the need for such an inventory.

Third, there should be in place some system of monitoring and reporting emissions. This is to ensure (among other purposes) that, if banking of permits were allowed, emissions permits to be sold by any sources would at least represent part of their real emissions reductions from the allowed emissions levels. This, combined with the above requirement for good emissions inventories, would provide certainty about the validity of permits traded, thus increasing confidence and incentives for inter-source trading.

Fourth, there should be effective enforcement aimed to detect those in non-compliance and to apply sanctions. Although enforcement is necessary for effective application of other instruments as well (e.g. charges and regulations), this requirement is of particular importance to emissions trading because, under an emissions-trading scheme, firms that operate in a country without adequate enforcement can emit without handing over their permits. Consequently, they can sell their permits to firms in other countries, thus leading to exceeded emissions in the sellers' country. By contrast, when charges or regulations are used, firms that defraud cannot sell permits to sources in other countries. Clearly, if enforcement were not adequate, it would be easy for a firm to sell permits or to refrain from buying permits without taking adequate measures to reduce its emissions. Consequently, an emissions-trading scheme would lead to higher overall pollution levels compared with instruments such as charges or regulations. Besides, enforcement at the international level often proves to be more difficult and less likely to be effective than at the national level because of the absence of an institution with the international jurisdiction to enforce policy. This further underlines the importance of national legal mechanisms for enforcement.

According to Article 17 of the Kyoto Protocol, the Parties with targets included in Annex B, which lists 38 countries and the European Community, may participate in emissions trading for the purpose of fulfilling their commitments under Article 3 of this Protocol. While the Article has a loose stance on the qualifying requirement as we propose, it indicates that emissions trading is limited to Annex B countries. Even with the scope of participation, given great differences in environmental monitoring and enforcement infrastructures among Annex B countries, however, it will probably take years to agree on the commonly accepted rules and guidelines "for verification, reporting and accountability for emissions trading" pursuant to the Kyoto Protocol. Even if they could have been worked out after lengthy negotiations, the scope of participation is very narrow, even in comparison with the WTO Members, who represent only part of the world community. Because non-Annex I countries have not committed themselves to any targets, pressure has been placed on Annex I countries to take trade measures to protect their domestic industries against competition from those countries that do not adopt GHG-emissions limits. If so, this will have far-reaching implications for the international trading system. This brings us to the next issue.

Some Trade-Related Aspects of Emissions Trading

The issue of compatibility of using trade measures against foreign environmental practices with the GATT was not given much attention until the findings of two GATT disputes panels on trade measures unilaterally taken by the United States in the US Mexico Tuna Dolphin disputes were made public. Both panel reports, which are commonly referred to as Tuna Dolphin I and Tuna Dolphin II, found the US restrictions on tuna imports from Mexico, which did not meet the US standards on dolphin-safe fishing practices, in violation of GATT. The panel in Tuna Dolphin I ruled that all trade restrictions directed against environmental harms have to be territorial. Moreover, such restrictions cannot be justified under Article III if they relate to the process of production rather than to the product itself. The panel explained that, if governments could regulate imports according to the production process by which they were made, the rules of the GATT's Article III would allow governments to require imports to conform to any type of social regulation currently imposed on the production process of domestic producers. It would allow governments to condition market access on compliance with domestic laws governing working conditions. The panel in Tuna Dolphin II concluded that Article XX does not preclude governments from pursuing environmental concerns outside their national territory, but such extra jurisdictional application of domestic laws would be permitted only if aimed primarily at having a conservation or protection effect. The second panel ruled that the US restrictions were in violation of GATT because they aimed to force other countries to change their own policies in order to comply with the US standards.

The preceding discussion promotes the concern about the compatibility of an international emissions-trading scheme with the GATT/ WTO. In what follows, I examine whether such an emissions trading scheme has the potential to bring parties into conflict with the WTO provisions in dealing with the allocation of permits, non-compliance with emissions targets, emissions-trading system enlargement, and trade measures against non-Members of an emissions-trading club.

Allocation of Permits

The Kyoto Protocol has set the caps on aggregate GHG emissions for Annex I countries. If trading among private parties is authorized, the next issue is how these governments allocate the assigned amounts within their countries. The allocation process itself represents the establishment and distribution of private property rights over emissions, and itself lies outside the mandate of the WTO.

The allocation of permits depends on the structure of national emissions-trading systems. Such systems could be modelled as either "upstream, " or "downstream, " or "hybrid" systems. An upstream trading system would target fossil-fuel producers and importers as regulated entities, so would reduce the number of allowance holders to oil refineries and importers, natural gas pipelines, natural gas processing plants, coal mines, and coal-processing plants. For example, if such a system were to be implemented in the United States, the total number of allowance holders would be restricted to about 1,900. Even with such a relatively small number of regulated sources, market power would not be an issue. In the above upstream system for the United States, the largest firm has only a 5.6 per cent market allowance share. Firms, with each having less than 1 per cent share, would hold the lion's share of allowances. Implemented effectively, an upstream system would capture virtually all fossil-fuel use and carbon emissions in a national economy. Firms would raise fuel prices to offset the additional cost. In an upstream system the number of firms that have to be monitored for compliance is relatively small and thus is easier to administer. Moreover, existing institutions for levying excises on fossil fuels, which exist in most industrialized countries, can be used to enforce the scheme. However, one of the drawbacks of an upstream system is that it provides no incentive for energy end-users to develop disposal technologies, the aspect that is deemed critical in seeking long-term solutions to solving climate change problems.

In contrast, a downstream trading system would be applied at the point of emissions. As such, a large number of diverse energy users are included. This would offer greater competition and stimulate more robust trading, thus leading to increased innovation. However, such a system would be more difficult to administer, especially concerning emissions from the transportation sector and other small sources. On the other hand, it would avoid the possibility that some energy users would not respond to the price signal, which might occur in an upstream system because of market imperfections such as high transaction costs, high discount rates, and imperfect information, although the extent of their responsiveness depends on the degree of competition and on whether price increases are actually passed on to the consumers. To keep a downstream trading system at a manageable level, regulated sources could be limited to utilities and large industrial sources. Governments could then address uncapped sources through other regulatory means such as carbon taxes. In doing so, however, the governments need to establish additional programmes. This would be an administrative burden and, in some countries, there would be great political difficulties involved in introducing carbon taxes. Moreover, the actual achievements in reductions of CO_2 emissions by a proposed carbon tax remain uncertain because of imperfect knowledge of the price elasticities of demand and supply for fossil fuels. This would put the governments at risk of non-compliance with the emissions commitments.

Alternatively, national trading systems could be modelled as hybrid systems. A hybrid system is similar to a downstream trading system in the sense that regulated sources at the levels of energy users are also limited to utilities and large industrial sources. On the other hand, like an upstream trading system, a hybrid system would require fuel distributors to hold allowances for small fuel users and to pass on their permit costs in a mark-up on the fuel price. As such, small fuel users are exempted from the necessity (and transaction costs) of holding allowances. Yet the rise in fuel price will motivate them to reduce fuel consumption or to switch from fuels with a high carbon content, such as coal, to fuels with a low carbon content, such as natural gas.

No matter how national trading systems are modelled, importers and domestic producers of fossil fuels should be treated equally in obtaining emissions allowances under the "like product" provisions in the WTO. Moreover, regardless of whether individual countries choose to empower private trading, the ultimate responsibility for fulfilling the Kyoto commitments would remain with the national government as a Party to the Protocol.

Given the great concern about international competitiveness, however, the allocation of permits does have the potential to bring Parties into conflict with the WTO provisions. Some fear, for example, that governments could allocate the permits in such a manner that domestic firms would be favoured over foreign rivals; this would violate the WTO principle of non-discrimination. The allocation of permits could also be designed in such a manner that certain sectors would have an advantage over others and further enhance their existing imperfect market competition. This makes the unequal treatment explicit, which can be much more easily hidden from the general public if the conventional command-and-control regulations are used. This, in turn, would have a price-distortion effect similar to that of a subsidy, and would be in conflict with the WTO rules that prohibit the use of export subsidies for such a purpose. All this clearly indicates that the manner in which countries allocate their assigned amounts should be compatible with these WTO principles and should not constitute a means of arbitrary or unjustifiable discrimination or a disguised restriction on international trade.

However, it should be pointed out that although grandfathering is thought of as giving implicit subsidies to some sectors, grandfathering is less trade-distorted than the exemptions from carbon taxes. To understand their difference, it is important to bear in mind that grandfathering itself also implies an opportunity cost for firms receiving permits: what matters here is not how you get your permits, but what you can sell them for that is what determines opportunity cost. Thus, relative prices of products will not be distorted to any great extent and switching of demands towards products of those firms whose permits are awarded gratis (the so-called substitution effect) will not be induced by grandfathering. In this way, grandfathering differs from the exemptions from carbon taxes. In the latter case, substitution effects exist. For example, the Commission of the European Communities (CEC) proposal for a mixed carbon and energy tax provides for exemptions for the six energy-intensive industries (i.e. iron and steel, non-ferrous metals, chemicals, cement, glass, and pulp and paper) from coverage of the CEC tax on grounds of competitiveness. This not only reduces the effectiveness of the CEC tax in achieving its objective of reducing CO_2 emissions but also causes those industries that are exempt from paying the CEC tax to improve their competitive position in relation to those industries that are not. There will therefore be some switching of demand towards the products of these energy-intensive industries, which is precisely the reaction that such a tax should avoid.

With the great concern that a government that grandfathers permits to a domestic firm could give it a competitive advantage over a similar firm in another country where permits are not awarded gratis, in the opinion of some there is a need for the harmonization of allocation of permits. However, I think that individual governments should be left free to devise their own ways of allocating permits on the following grounds. First, I think this is not necessarily the case, because even if a firm obtains emissions permits by auction, its government still can, if necessary, protect its international competitiveness by means of recycling the revenues raised through auctioned permits to lower other pre-existing distortionary taxes, such as taxes on labour and capital.

Second, although auctioning at least part of the assigned amounts to subnational legal entities alleviates to some extent the concern about international competitiveness, any attempts to produce an agreement on a common rate are likely to run into concerns about national sovereignty and thus would encounter significant political difficulties. Take the above CEC proposal for a carbon energy tax as an example. National sovereignty considerations to some extent explain why the CEC proposal for a carbon energy tax failed to gain the unanimous support of its member states, partly because some member states opposed an increase in the fiscal competence of the Community and thus opposed the introduction at a European level of a new tax on grounds of fiscal sovereignty.

Third, given great differences in national circumstances, setting a uniform rule of allocation will restrict the rights of individual governments to select the option that is best suited to their own national circumstances. Indeed, the failure of the above CEC carbon energy tax is to some extent because some member states are loath to restrict themselves to the common CEC-specified policy-and-measure design to stabilize CO_2 emission and how to do it. It is conceivable that some countries whose economies are heavily distorted would decide to auction permits, and that the revenues generated through auctioned permits could then be used to reduce pre-existing distortionary taxes, thus generating overall efficiency gains. Parry, for example, show that the costs of reducing US carbon emissions by 10 per cent are four times more under a grandfathered carbon permits case than under an auctioned case. This disadvantage reflects the inability to make use of the revenue-recycling effect in the former case.

Fourth, and importantly, leaving individual governments the freedom to devise their own ways of allocating assigned amounts to subnational entities would ensure that any individual government maintains its right to determine the domestic policies and measures that would be taken to meet its Kyoto obligations. For example, a government that wants to use taxes or regulations for domestic emissions control could retain the sole right to trade. Alternatively, a government could allocate its assigned amounts to private entities to trade.

Emissions-Trading System Enlargement

The Kyoto Protocol allows Russia to emit the same volume of GHGs as (the then) Russia did in 1990. Given the fact that CO_2 emissions in Russia declined over the past years after the collapse of the Soviet Union in 1991 and are expected to continue to decline, that means that Russia should be left as the biggest seller of emissions permits among Annex I countries once emissions trading takes place. The United States has reached a conceptual agreement with Australia, Canada, Japan, New Zealand, Russia, and Ukraine to pursue an umbrella group to trade emissions permits. It is believed that the United States is counting on emissions trading with Russia to achieve half of its 7 per cent reduction target set in the Kyoto Protocol. Although the United States insists on bringing non-Annex I countries into an emissions-trading scheme, which has widely been seen as creating a source of such permits for the United States to buy and therefore achieve its own agreed reductions through offshore compliance, it might seem that the United States does not want the EU to breathe Russian "hot air" because the addition of the EU to the group would raise the prices of the Russian emissions permits that the American firms would have to pay. On the other hand, Russia would not welcome the addition of non Annex I countries, such as China and India, to the group because these new entrants would raise the supply of overall permits on the market and depress the prices of those permits held by Russia. This is also one of the reasons for the developing countries' opposition to emissions trading because they feel that it leaves them out of the system at this stage, although for some legitimate reasons these countries do not want to join in an emissions-trading club at this moment. Although these are just political speculations, they underline the importance of establishing clear rules of procedure about admitting new entrants before emissions trading begins.

There are two avenues to establish such rules of procedure. One is based on voting to admit new entrants. So far, any decisions made by the Conference of the Parties to the UNFCCC have been generally adopted by consensus. If admitting new entrants requires consensus by all current Annex B countries eligible for emissions trading, this confers on Russia a de facto power of veto. Thus, if the avenue to admit new entrants rests on voting, a three-quarters majority vote of the current Annex B countries present and voting at the meeting could be adopted to prevent exploitation of market power.

The second avenue rests on automatic phase-in once one prospective country meets predetermined criteria. In my view, the second avenue is superior to the first. Such criteria should include under what conditions, how, and when a new entrant could be incorporated into the emissions-trading scheme. Once such criteria are set, they should remain stable in the short run although this by no means precludes any adjustments that might be required in the future.

Similar reasons hold for expansion to include GHGs other than CO_2 and the uptake of GHG by sinks, because a comprehensive coverage of both gases and options will induce more cost-effective abatement options. On the other hand, a workable emissions-trading scheme requires that emissions of whatever pollutant is to be included have to be measured with reasonable accuracy. This requirement implicitly precludes including all gases in the initial trading scheme. However, limiting trading to a subset of gases is not likely to be effective unless the Protocol is further amended to partition the assigned amounts into two categories tradeable and non-tradeable gases with separate goals being assigned for each. Without a separation of categories, there appears to be a lack of any legal basis for rejection of the legitimate claim from those countries that use the flexibility inherent in the equivalence process to substitute freely among the gases, because Article 5.3 of the Protocol has authorized that the global-warming potentials are used to translate non-CO2

GHGs into carbon-equivalent units in determining each Annex I party's compliance with its assigned amounts.

Non-compliance with the agreed emissions targets

The Kyoto Protocol itself does not contain stipulations on what actions, if any, should be taken in the event that a country were found to be in non-compliance. Without clear criteria to judge compliance, it is difficult to believe that many of the Annex I countries will be willing to ratify the Protocol. The United States had proposed penalizing a country that failed to meet its target by imposing ever larger reduction obligations over a subsequent period (the so-called borrowing with a penalty); however, negotiators at Kyoto blocked the only compliance mechanism on the table because they fear that it would not bring any additional pressure to bear on a country that simply continues to disregard its commitments. It is then natural to consider the use of trade measures usually in the form of a trade restriction to enforce a country to comply with its commitment. In this case, caution should be taken because all WTO retaliation is limited to a "compensatory" amount that is, an amount equivalent to the value of the trade obligations being nullified or impaired by the other Party. Its main significance is that it rejects a more aggressive approach towards sanctions the approach under which a legal sanction must be large enough to produce the desired change of behaviour.

Trade measures against non-members of an emissions trading club

Because non-Annex I countries have not committed themselves to any targets, Annex I countries have been pressed by their powerful lobbying groups to take trade measures to protect their domestic industries against competition from non-Annex I countries. What kind of trade measures could potentially be applied against nonmembers of an emissions-trading club in order to counter the trade effects of honouring their Kyoto commitments? One possibility might be that Annex I countries comply with their Kyoto commitments, but use border adjustment taxes based on embodied carbon content of goods imported from non-Annex I countries to keep non Annex I countries' emissions at their baseline levels. Another possibility is that Annex I countries abide by their Kyoto commitments, but set imports of energy-intensive goods from non-Annex I countries at their baseline levels or set exports of energy-intensive goods from non-Annex I countries at their baseline levels. The third possibility is that Annex I countries abide by their Kyoto commitments but provide subsidies up to the levels at which their exports of energy-intensive goods remain at their baseline levels. Although these trade measures appear to include only the component designed to reduce GHG-emissions leakage, they are, in principle, in conflict with the GATT/WTO principles of most-favoured nation and nondiscrimination. Such taxes could fall foul of the "like product" provisions in the GATT's Article I and Article III, which are designed to prevent a country from discriminating against imports on the basis of their territorial origin. There are formidable technical difficulties involved in identifying (and it may even be completely impossible) the appropriate carbon contents embodied in virtually all traded products unless non-Annex I exporting countries are willing to cooperate in certifying how the products are produced. In the absence of any information regarding the carbon content of the products from non-Annex I countries, importing Annex I countries may prescribe the tax rates based on their domestically predominant method of production for the imported products. Such a practice will violate the GATT rules that do not allow trade measures to be taken on a basis of the differences in process and production methods (PPM), and appears to deprive non-Annex I countries of enjoying the very basis of comparative advantage in their production. Moreover, such tariffs would be likely to violate commitments made by the WTO contracting countries not to raise import taxes above "bound tariff" levels, i.e. maximum tariffs for goods listed in an annex to the GATT. The WTO rules also prohibit the use of export subsidies to give certain sectors an advantage over others.

Are there any potential avenues within the WTO rules to accommodate Annex I countries on this score, should they decide to pursue the above offsetting trade measures? One avenue would be that Annex I countries could claim general exceptions under the GATT's Article XX, which states that:

Subject to the requirement that such measures are not applied in a manner which would constitute a means of arbitrary or unjustifiable discrimination between countries where the same conditions prevail, or a disguised restriction on international trade, nothing in this Agreement shall be constructed to prevent the adoption or enforcement by any contracting party of measures:

… (b) necessary to protect human, animal or plant life or health;

… (g) relating to the conservation of exhaustible natural resources if such measures are made effective in conjunction with restrictions on domestic production or consumption;

Article XX itself sets out the exceptions that authorize governments to employ otherwise GATT-illegal measures when such measures are necessary to deal with certain enumerated social policy problems. Since Annex I countries participating in emissions trading are obligated to limit their GHG emissions, whereas non-Annex I countries are not bound by such commitments, the same conditions do not exist for Annex I countries and non-Annex I countries. Therefore, Annex I countries could argue that the discrimination in the trade restrictions is neither arbitrary nor unjustifiable, thus meeting the requirement in the preamble to Article XX. Under Article 2.2 of the Agreement on Technical Barriers to Trade, technical regulations shall not be more trade restrictive than necessary to fulfill a "legitimate objective," which is defined as including the protection of human health or safety, animal or plant life or health, or the environment; the offsetting trade measures could be based on a legitimate environmental objective and not merely on formal membership of an international agreement. Thus, if any non-Annex I countries voluntarily assume binding emissions targets (a prerequisite for engaging in emissions trading, through amendment to the Annex B of the Kyoto Protocol), they should be exempted from such trade restrictions. This is of particular importance to newly industrializing countries because they are the countries that are most likely to make such voluntary commitments among non-Annex I countries. This is not without precedent: for example, the Montreal Protocol has included the provision that exempts non-Parties from trade measures if they are determined by the Parties to be in compliance with the phase-out schedules. Indeed, the United States has proposed at Kyoto to allow non-Annex I countries to adopt GHG controls voluntarily so that non-Annex I countries can also be brought into an emissions-trading club. However, the group of 77 and China blocked the US proposal. The underlying reasons for their objections are given below. First, on ethical grounds, non-Annex I countries think that Annex I countries have caused the climate-change problem and should thus clean it up themselves before asking non-Annex I countries to take action. Second, non-Annex I countries insist that the US demand for developing countries' commitments goes against an earlier UN agreement (known as the Berlin Mandate) adopted at the first Conference of the Parties to the UNFCCC in Berlin in April 1995, which specifically indicated that "a protocol or another legal instrument" adopted at the third Conference of the Parties in Kyoto should "not introduce any new commitments" for non-Annex I countries. Third, there is the widespread fear that the opportunity to trade emissions permits might eventually lure others in the group, especially the Latin American nations, to be drawn into making commitments of their own. Then rich Annex I countries might use emissions trading as a means of buying their way out of responsibility for climate problems and at the same time postponing the radical changes in their own consumption patterns and passing the responsibility on to the poor, while GHG-emissions limits grow subsequently more restrictive for the developing countries.

In addition, Annex I countries might argue that to support the operation of an emissions-trading scheme that is designed to protect health and conserve depletable fossil fuels, some kind of offsetting trade measures would be required and qualify for the exceptions under Article XX (b) and Article XX (g) because international emissions leakage would otherwise frustrate its intent. Furthermore, although such measures would not be the least trade distorting, they are much less trade restrictive than the tariff suggested by Hoel, which also includes the optimal trade tariff term.

Another avenue rests on Article 1.1 of the Agreement on Subsidies and Countervailing Measures, under which a subsidy is defined as including "government revenue that is otherwise due is foregone or not collected". Although a narrow interpretation of this clause would limit claims to cases in which taxes are levied but not collected, its broad interpretation would expose the absence of environmental taxes to charges of unfair subsidization. Annex I countries could argue, therefore, that the absence of emissions targets for non-Annex I countries would be equivalent to giving an implicit unfair export subsidy biased towards their energy intensive sectors (the so-called ecological dumping), because the costs of environmental degradation are not part of the prices of those exported products.

Although Annex I countries could justify their offsetting trade measures according to Article XX and the definition of lax environmental regulation as a counter available subsidy, non-Annex I countries of the WTO would, no doubt, resist such interpretations. As environmental issue was simply not a public issue in 1947 when GATT was signed, exploring the environmental exceptions to the general free-trade requirements in GATT depends as much on interpretation as on the actual clauses. Given the fact that a three-quarters vote of the entire membership is required for the membership to adopt legal interpretation of any WTO agreement, the above interpretations might stand little chance of being accepted by non-Annex I countries that form the majority of the WTO Members.

In addition, lessons from the Montreal Protocol may be instructive. The Protocol prohibits trade with non-Parties in ozone-depleting substances (ODS) themselves, products containing ODS, and possibly products made using (but not containing) ODS. The trade restrictions were used together with financial assistance (i.e. Multilateral Fund in this case) and technology transfer as a means to coerce or force countries to become Parties. Although the Protocol has succeeded in securing universal participation by means of the "carrot and stick" approach as the signatories to it make up more than 95 per cent of current world consumption and production of ODS, the fact remains that restrictions are discriminatory. Although the restrictions have not been contested by the WTO Members, the WTO Secretariat has voiced its opposition to such uses of trade restrictions, and its CTE has opted not to welcome their replication in an emissions-trading scheme, because such measures appear to violate the GATT/WTO principles of most-favoured nation, national treatment, and non-discrimination. In the Montreal Protocol, these discriminatory restrictions are not that important (given the scope of its membership, which is wider than that of the WTO itself), since most of the potential for trade conflict arises with non-Parties to the Protocol. However, if such restrictions were used in climate problems, the economic implications could be substantial because an emissions-trading club is only a small subset of the WTO Members at least at its initial stage. Moreover, there may be some legitimate reasons why many non Annex I countries, if not all, might want to remain outside an emissions-trading club unless side payments are offered to these countries so that they are not made worse off as Members than they would be as non-Members. Thus, it seems unfair to discriminate against them on merely this basis. Is there, then, any way out? This leads to another major subject.

Joint implementation with developing countries

Countries may differ with respect to assimilative capacities and to tastes and preferences for environmental quality. Thus, the harmonization of environmental policy would not be always necessary from an environmental point of view, and may lead to sub-optimal use of the environment from an economics point of view. On the other hand, from the industrialized countries' perspective, the lack of developing countries' involvement in combating climate change aggravates their short-term concerns about international competitiveness. Non-participation by developing countries increases emissions leakage that could arise in the short term, as emissions controls lower world fossil-fuel prices and, in the long term, as industries relocate to developing countries to avoid emissions controls at home. In addition, it raises the spectre of developing countries becoming "locked in" to more fossil-fuel-intensive economy and eliminates the Annex I countries' opportunity to obtain low-cost abatement options. Testifying on 4 March 1998 before the House of Commerce Subcommittee on Energy and Power about the domestic economic implications of the Kyoto Protocol, Dr Janet Yellen, Chair of the White House's Council of Economic Advisers, stated that when there is no emissions trading at all, the cost of complying with the Kyoto target for the United States would run to US$125 per ton of carbon. With emissions trading only among Annex I countries, the cost would drop to $30–50/ton. With fully worldwide emissions trading, the cost would further drop to $14–23 per ton. This clearly explains why the United States puts heavy emphasis on the involvement of developing countries. Indeed, recent Indonesian bush fires choking South-East Asia served as a graphic reminder that developing countries have an important part to play in protecting the environment against global warming.

For some time to come, however, many developing countries would not be qualified to participate in an international emissions trading scheme. This promotes the concern: how can we encourage their participation in combating global climate change, given the fact that there are a great deal of low-cost abatement options there? One widely recognized option to bring the developing countries on board is by means of JI. Indeed, many countries in the Organization for Economic Co-operation and Development (OECD) are keen to see JI as a key part of the Kyoto Protocol, even although it is not without conceptual and operational problems such as the form of JI, criteria for JI, the establishment of baselines against which the effects of JI projects can be measured, and the verification of emissions reductions of JI projects. In brief, JI means that the investor in one country invests in emissions-abatement projects in another (host) country where the costs of abating GHG emissions are lower than those involved when trying to achieve an equivalent degree of abatement within the home country and is credited, in whole or in part, for emissions abatements in its own GHG accounts. By shifting the burden of carrying out abatement in non-Annex I countries, JI thus offers the potential for lowering the global costs of abating GHG emissions and succeeds in reducing GHG-emissions leakage without discriminating against such countries. In the WTO jargon, JI offers a positive incentive for countries to participate in an international agreement:

When cooperation is not voluntarily forthcoming, positive incentives are the best way to achieve sustained inter-governmental cooperation. Positive incentives can include offers of financial assistance and transfers of environmentally friendly technology directly related to the problem at hand, as well as more broadly based offers, for example, to increase foreign aid, to lessen debt problems and to make non-discriminatory reductions in trade barriers.

However, bringing the developing countries on board and integrating JI credit trading into an international GHG-emissions scheme promotes another concern about the credibility of the scheme. How can such an international scheme incorporate credits from JI projects and at the same time ensure that confidence in the scheme is not compromised? One option would restrict the amount of JI credits that could be bought by Annex I countries for compliance from nonAnnex I countries. Another option, which is superior to the first option, would be to discount the credits awarded to JI projects. Such reduced crediting could provide an "environmental bonus" and, at the margin, allow for the uncertainty about reported emission reductions of JI projects. Moreover, I advocate a predetermined discount factor, a view shared by the Center for Clear Air Policy. In order to reflect the characteristics of a JI project and the differing quality of GHG monitoring and reporting infrastructures across countries, such a discount factor should differ according to both type of project and country and should be accordingly adjusted over time for those countries in order to reflect the improvement in their monitoring and reporting infrastructures. I think that a predetermined discounting approach is superior to a market-driven discounting approach, because the former would protect against the introduction of false credits into an emissions-trading scheme and provide non-Annex I countries with financial incentives to opt for binding commitments and develop their stringent monitoring and reporting infrastructures. In addition, only verified and certified JI credits would become part of the emissions-trading scheme. Once certified, these credits could be treated as equivalent in quality to all other permits.

It should be pointed out that the group of 77 and China have not proved very receptive to the concept of JI, for some legitimate reasons. It is also unclear if the US Congress will support such a mechanism, although for reasons very different to those used for the group of 77 and China. Given the fact that JI, by definition, will not shift the burden of paying for abating GHG emissions into non-Annex I countries, which could result in a large outflow of investment capital from the United States to non-Annex I countries, some influential congressmen in the United States have already begun to regard JI as foreign aid.

Summary

The Kyoto Protocol, despite its apparent flaws in its current form, is the first international environmental agreement that sets legally binding GHG-emissions targets and timetables for Annex I countries. Its Article 17 authorizes emissions trading between Annex I countries. If properly designed, emissions trading can effectively reduce these countries' abatement costs while assisting them to achieve their Kyoto obligations.

This chapter has examined the compatibility of an international emissions-trading scheme with the GATT/WTO. It has been found that, in dealing with the allocation of permits, with non-compliance with emissions targets, and with emissions trading-system enlargement, emissions trading has the potential to bring Parties into conflict with the WTO provisions. WTO will also have to resolve the very difficult question of what to do about Annex I countries of the WTO, should they decide to pursue the offsetting trade measures against non-Annex I countries of the WTO, who for some legitimate reasons have decided to remain outside an emissions-trading club. In this regard, JI may offer the way out because it succeeds in reducing GHG-emissions leakage without discriminating against such countries. However, the prospect for JI depends on how JI is implemented and on whether mutual mistrust between Annex I and nonAnnex I countries can be removed. In any case, breakthroughs in low-cost energy-efficient technologies and the ways to transfer such technologies play a key role in acceptably drawing developing countries into the battle against global climate change. The ozone experience has shown that, when low-cost substitution technologies start to become available and when their transfers to developing countries take place on fair and most favourable conditions, there is little difficulty in persuading developing countries to join. This underlines the importance for governments to enhance energy R&D and for the WTO Committee on Trade and Environment to explore the possibility of envisioning a more flexible patenting and intellectual property rights scheme that allows developing countries to acquire such technologies on preferential terms under the Agreement on Trade-Related Aspects of Intellectual Property Rights.

It should be pointed out that, although GHGs offer an even more attractive case for application of emissions trading than many local pollutants already well handled with emissions trading, the US SO_2 Allowance Trading System cannot merely be transplanted to the international terrain where legal and institutional ingredients differ substantially from those in the United States. It will probably take years, if not decades, to set up such an international scheme. This raises the political question about the credibility of achieving Kyoto commitments and, at the same time, promotes the necessity of investigating effective national policies that can influence the behaviour of firms and consumers and establish the credibility of long term goals. As such, interim measures before the beginning of the first commitment period in 2008 warrant special attention. They include national emissions-trading schemes with looser controls than are required by the Kyoto Protocol, crediting early emissions reductions below internationally accepted national baselines prior to 2008 within the jurisdiction of the advanced OECD countries, tax incentives to promote energy-efficient technologies, signals of carbon taxes to be levied at a specific future date, and government's role as a larger buyer to create stronger demand for energy-efficient technologies.

7

Role of Computers in Global Warming

Introduction

When life first appeared on our planet, whether by chance or design, it was possible only because the climate was friendly; it was neither too hot nor too cold. The atmosphere, containing water vapour, carbon dioxide, and other components, acted like the glass panels of a greenhouse, letting the sun's heat in but preventing some of it from escaping, thus warming the planet. If it were not for this greenhouse effect, temperatures at the earth's surface would be far colder than they are, and life as we know it could not exist.Global warming has become an issue of concern because of the perception that increasing greenhouse gases will cause the earth to warm so fast that nature (and humankind) may not be able to adapt to the rapid change. The atmosphere has always acted as a warming blanket around the earth. It is not the warmth of this blanket but the possibility of an abrupt or extreme temperature change that is causing some scientists concern.The popular view of global warming—the "myth"—has developed over the past decade because of predictions of computer climate models. This view has been presented in books by authors such as Al Gore, Stephen Schneider, Isaac Asimov, and Paul Ehrlich, and it has been advanced by the media and in school classrooms. Environmental organizations have taken up the cause and demanded political action. At times the reaction to this view reaches a feverish pitch, generating hysteria. Chapter 2 covers this phenomenon and examines some reasons for it.What is the global warming myth? It is a scenario of doom that contains certain facts and also certain assumptions:

- Atmospheric carbon dioxide (CO_2) causes warming of the planet.
- Man's activities are increasing the amount of carbon dioxide.
- The average temperature of the earth has increased approximately 0.5°C (0.9°F) in the last 100 years.
- Global temperature will increase another 1.5-4.5°C (2.7- 8.1°F) by sometime in the next century if we do not take drastic measures.
- The predicted results of this warming include melting of the polar ice caps, flooding of coastal cities, massive extinction of species, and the deterioration of civilization as we know it.

What is the basis for the myth? Which components of the myth are proven facts, and which are assumptions based on computer modeling?

Global Warming: Background

The issue of global warming first emerged when scientists became aware of the amount of carbon dioxide (CO_2) being added to the atmosphere as a result of human activity. However, the first scientists to recognize the relationship between carbon dioxide and climate were concerned about excessive cooling, not warming.

Jean-Baptiste Joseph Fourier, a French scientist during the Napoleonic years, is generally given credit for being the first to describe the greenhouse effect in the 1820s. Fourier was known for both his developments in mathematics and his studies of Egypt. He studied the properties of heat its radiation and transfer and he developed the mathematics of partial differential equations. His discovery of the greenhouse effect was a result of his heat studies.

John Tyndall, another scientist interested in the greenhouse gas phenomenon, was an Irish physicist and Professor of Natural Philosophy at the Royal Institution in London. (Tyndall is best known for his studies of light scattering from the earth's atmosphere, which led to our understanding of why the sky is blue.) In the 1860s,

Tyndall measured the radiation absorption efficiencies of various gases, a measure of their effectiveness as greenhouse gases. He was concerned that a decrease in atmospheric CO_2 could lead to another ice age.

In 1896 Svante August Arrhenius theorized that carbon dioxide was a greenhouse gas being introduced into the atmosphere by the burning of carbon-based fossil fuels. Unlike some of today's scientists, he concluded that any warming caused by this effect was good for the human race. He looked forward to a warmer climate that would bring more abundant crops "for the benefit of rapidly propagating mankind."[1] At the time, no one got particularly excited about his global warming theories. Arrhenius, who taught himself to read at the age of three, did not always operate within the popular framework of scientific research, being accused of, among other things, manipulating imaginary data. Ultimately, many of his ideas began to be accepted. In 1903 he became the first Swede to win the Nobel prize, and he is now considered the founder of physical chemistry. Even so, some of his theories remain outside the realm of acceptance, such as his suggestion that life is spread by bacteria activated by collisions of stars.

Little was written about carbon dioxide as a greenhouse gas for the next forty years. In 1938 G. S. Callendar, an English meteorologist, evaluated historical records of atmospheric carbon dioxide and showed a trend of increasing concentration.[2] He, too, welcomed the potential future warmth and felt that the carbon dioxide would provide fertilizer for increased farm production. He hoped that the return of the deadly glaciers would be delayed indefinitely.

Eventually, climatologists became concerned that the increase in carbon dioxide would cause too much warming. In 1956, Gilbert Plass, a scientist at Johns Hopkins University, Baltimore, Maryland, expanded on a theory that attributed great importance to carbon dioxide. He suggested that atmospheric carbon dioxide controls the climate, and he predicted that accumulated atmospheric carbon dioxide from mankind's burning of fossil fuels would cause a global temperature increase of 1.1°C (2.0°F) by the beginning of the twenty-first century.[3] It was well known by then that large quantities of carbon dioxide were being emitted into the atmosphere because of the fuel used to meet the energy and industrial requirements of society.

Increases in atmospheric carbon dioxide were actually observed about this time by various scientists working independently of each other. The most prominent was Charles David Keeling of Scripps Institution of Oceanography, La Jolla, California. Keeling measured the atmospheric carbon dioxide concentration at various locations in the United States and found that it had increased from about 275 parts per million (ppm) in the nineteenth century to 311 ppm in 1956.

The warnings became dire (and much more publicized) as Syukuro Manabe and Richard Wetherald, climatologists working at Princeton University, predicted increases of about 2°C (3.6°F) by the end of the twenty-first century. This prediction was based on one of the first of many modern complex computer models. Ironically, this study was reported during a time when many scientists, observing a global cooling trend from about 1950 until 1980, were predicting the beginning of another ice age.

During the 1980s, computer modeling of the earth's climate and the effect of doubling atmospheric carbon dioxide became increasingly popular among climatologists. This modeling activity and heavy speculation concerning the consequences of the predicted warming resulted in enhanced coverage by the scientific press, including the weekly general science journals such as Science and Nature.

> *Publicity about the imminent danger of global warming was brought to a head in 1988, one of the hottest years on record, when James Hansen, director of NASA's Goddard Institute of Space Studies in New York City, testified before Congress that global warming had begun! In 1989 he testified again and defended his testimony in an interview with Richard Kerr, a research newswriter for Science, stating:*
>
> *I said three things. The first was that I believed the earth was getting warmer and I could say that with 99% confidence. The second was that with a high degree of confidence we could associate the warming and the greenhouse effect. The third was that in our climate model, by the late 1980s and early 1990s, there's already a noticeable increase in the frequency of drought.*

His statements were so strong that the Office of Management and Budget required that they be qualified with a statement about the uncertainties of computer models.

At the time of Hansen's 1989 congressional testimony, the Workshop on Greenhouse-Gas Induced Climatic Change was being held in Amherst, Massachusetts. Most of the climatologists at the workshop felt that Hansen had gone too far, overstepping the boundaries of good science with his pronouncements. The forty participants who were present on the last day of the workshop issued a press release that stated: It is tempting to attribute [the 0.5°C warming of the past 100 years] to the increase in greenhouse gases. Because of the natural variation of temperature, however, such an attribution cannot now be made with any degree of confidence.

Scientists rarely like to state uncertain events in absolute terms, and it is not surprising that this group of peers was upset with Hansen for publicly stating speculation as fact.

Even Stephen Schneider, a prominent climate modeler at the National Center for Atmospheric Research and proponent of the global warming theory, pointed out the weakness of Hansen's position when he stated, "He's not running a realistic ocean." This is a reference to the computer model of the ocean that is incorporated into the climate model. Schneider went on to say, "They [Hansen's group] have been using a pretty hokey ocean, we all have. But you have to have less confidence because of that." In spite of these statements, Schneider still believes that global warming is beginning and will cause major problems in the future.

We focus on the computer climate models and explain how they work, discuss the uncertainties associated with their input data, and evaluate their predictions. There we will discuss the "hokey ocean" and other simplifications that are common practice in computer modeling.

There is little controversy over the greenhouse effect as a scientific theory. The controversy arises over the global warming predictions: the causes (whether or not they're human induced), the amount of warming and the timing of it, the implications for life and society, and most of all, what to do about it.

Let us take a more detailed look at the greenhouse effect, global warming, and scientific prediction.

Greenhouse Effect

The earth derives most of its energy from the sun, which continuously emits radiant energy equivalent to a 6,000°C (about 11,000°F) boiling caldron. This radiant energy, a very small portion of which reaches our stratosphere, consists of the full range of radiation, including the ultraviolet (UV), visible, and infrared (IR) portions of the sun's spectrum. On the one hand, ultraviolet radiation, the most energetic portion, causes damage to living tissue if it is overexposed; it is UV radiation that causes sunburn and blinds people who stare directly at the sun. On the other hand, the interaction of UV and visible radiation within plants provides the energy required for the photosynthesis reactions essential to plant growth. Since plants are the foundation of our food chain, this radiation is essential for all life.

The visible portion of the sun's radiation is simply that to which our eyes are sensitive. We cannot see the UV or IR portions of the sun's spectrum, but we can feel the warmth of IR radiation it is the same as that which is radiated from an infrared heat lamp. The IR energy is too low to trigger our optical nerve endings, so our eyes are not sensitive to this radiation.

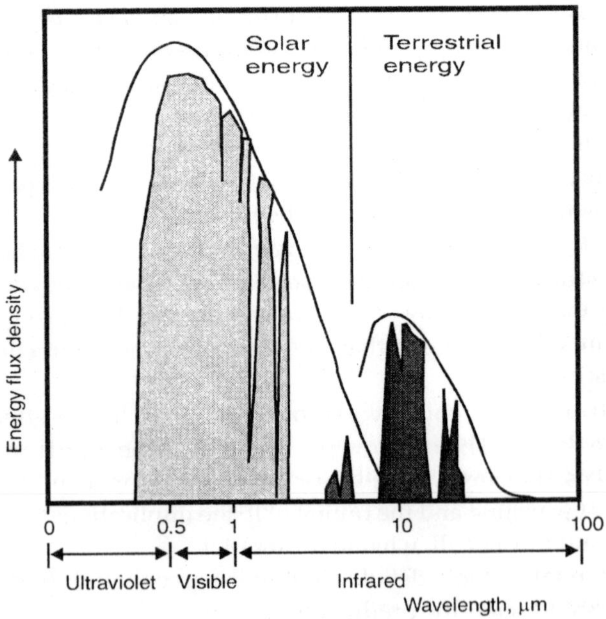

The sun furnishes the earth with a generous amount of ultraviolet, visible, and infrared radiation because it radiates at such a high temperature. The earth in turn radiates some of the sun's energy back into space at a very low temperature, about 15°C (59°F).[9] Because of this low temperature, nearly all of the earth's radiation is in the low-energy infrared region.

If there is a balance between the energy provided by the sun and that returned to space by the earth, the earth's temperature will remain constant. If the earth retains more energy than it returns to space, it will heat up; conversely, if the earth retains less energy than is returned to space, it will cool down.

When the sun's radiation enters the atmosphere, many things can happen to it. When sunlight encounters dust or clouds or any other type of matter, it may be absorbed, reflected, scattered, or transmitted. Some of the radiation passes through to the earth; some of it is reflected back into space; some of it interacts with oceans or plants and is absorbed; some is scattered or reflected from the ice and snow cover.

While solar radiation can interact with all matter, how it acts depends on the portion of the solar spectrum to which it belongs: UV radiation has higher energy than IR radiation, and visible radiation is in between. A good example of this interaction occurs in the ozone layer in our stratosphere. The ozone layer is very efficient at absorbing very high-energy UV radiation, but it lets much of the lower-energy UV and most of the visible and IR radiation pass right through into the lower atmosphere, or troposphere. This radiation ultimately interacts with clouds or dust in the troposphere or with the earth itself, causing the earth to heat up proportionally to the energy that is transferred.

Finally, there is the skill we probably look for most of all in leaders: authenticity. Everyone talks about charisma, but the fact is that an infinitesimally small percentage of the population are universally acknowledged to possess that quality. Among senior executives, real charisma stems from authenticity: who that person is when you talk with him/her versus when he/she is in the boardroom or in front of customers. An authentic, charismatic business leader is someone who is consistent and not a sort of chameleon, changing colors for certain groups. You get strong credibility if you establish yourself as the genuine article. People will follow a leader who is authentic and passionate, as management gurus Jim Collins and John Kotter attest in their interviews in this chapter.

While the general attributes of the new-style leader are consistent across industries and organizations, the sizes and developmental stages of various organizations may call for different combinations of competencies. Skills that delineate to early-stage companies, for example, include extraordinary amounts of energy and intensity, the ability to multitask, an unusually high tolerance for ambiguity, and the aforementioned agility. Typically, a company in this phase has only a few products or services or is building a broader service offering to one or two product lines, so every step counts and there is little room for a misstep. Orbitz's CEO, Jeffrey Katz, had the formidable task of launching a start-up dotcom company after the dotcom crash in 2000. He acknowledges that, faced with these pressures, "it is essential to be a good strategist and planner: to set goals, to build clarity and consensus around them, and to connect the short-term goals with long-term results through excellent execution. Meeting a series of short-term goals helps to bring about positive and lasting changes. " In my experience, the most successful executives are also those who can morph to the market conditions, be they changing capital requirements or capital supply. Being the CEO of any company is difficult, but being the CEO of a start-up is like throwing yourself in front of a train every day.

At the midcap level, there is a change in the required mix, moving from absolute personal leadership toward the competencies involved in empowering the team, or teams. It is an issue of scope and scale for a rapidly growing company. You have to delegate very effectively and empower business units as the organization grows. Also, leaders here have to continue to be passionate about key objectives, meticulously examine the business, and put a focus on change management. Executives with these abilities are likely to have fun in this environment and to succeed in creating breakthrough organizations. Given today's burgeoning new technologies, a company's best practices can be copied very quickly. Competitors can emulate what another company accomplishes within weeks or months. They may not be as good at it, but in the process, your company may have lost its ability to differentiate its products and services in the marketplace; that is why the ability to stay ahead of the change curve is so critically important in madcap companies.

The executives leading a multinational company are more in the league of senior diplomats than CEOs; the demands are transcendent of traditional leadership skill sets and call for the ability to synthesize experience as much as to analyze it. Theirs is the realm of strategic alliances, industry exchanges, and information portals. Their stature is such, their EQ is at a level, that it is impossible to articulate, let alone emulate, how they do what they do. At this level, the ability to step beyond the personal and see the world and one's role in it with objectivity, compassion, and resolve is critical. These types of leaders are genuine and confident of their abilities. They have a statesmanlike approach to representing a company, and it is a learning approach in which they acknowledge and examine mistakes and opportunities not taken but still celebrate success. Michael Jordan claims that one of the ways he became such a great basketball player is by missing over nine thousand shots in his professional career. That says it all.

For companies of all sizes, the implications of the new demands and stresses on their leaders seem crystal clear: in the realm of human capital management, the time has come to translate lip service into action. The leaders featured in this chapter, Linda Sanford at IBM, Hank McKinnell at Pfizer, Fred Smith at FedEx, and David Pottruck at Charles Schwab, were among the first to turn talk into action by empowerment, by strategic recruiting, by investing heavily in management development, and in a myriad other ways that are evident without being easy to define.

Also inherently important for a company of any stage or size is the capacity for change management and corporate transformation on an almost daily basis.Greg Owens at Manugistics Group developed a people strategy concurrently with a turnaround strategy. He notes: "This series of strategies is ongoing, because to maintain success, an organization must constantly work at recruiting, developing, and nurturing its intangible assets. " This emphasis on people has made the difference between success and failure, Owens believes.

Recruiting is tougher all the time, as the supply of high-potential managers continues to dwindle, and even landing the best of the lot is not adequate unless you can hang onto them in a very competitive marketplace for talent. The interviews that follow have a lot of excellent guidance on "best practices" in recruiting and retaining talent. And, interestingly, not all of them have to do with financial incentives. Look at WWF International, for example. Thanks to its mission and culture, notes director general, Claude Martin, the conservation organization rarely has to recruit. Candidates flock to the organization, wanting to serve, and, once hired, rarely leave.

It's something you can just feel, walking into a company where human capital is highly valued. It is palpable in the interviews in the following pages. A handful of others, such as Dell Computer and Cisco Systems, have figured it out, too. Changing the rules and making them stick in the marketplace investing time, money, and talent into human development even though it will take years to see the results takes courage and commitment. So when you see a company do it, it is very impressive. The companies that don't do it will soon wish they had.

Leadership by courage, creativity and integrity

- What are some of the defining moments in your career that have shaped your leadership style?

  I describe myself as an intellectually curious person who likes to take on a variety of challenges. I'm very interested in taking what I learn about different areas and applying those lessons to the things I am involved in to make an impact. The first defining moment for me was the decision to work at Bain & Company. While my stay there was just shy of three years, it had a major influence on me by exposing me to a range of industries. It also gave me an appreciation for how to identify and focus on the key issues that drive a business. I also learned the importance of challenging conventional wisdom. As I moved into the corporate world, leaving Bain for American Express, I learned that leadership comes in different forms, in different opportunities. One of my first assignments was in strategic planning, where I had to evaluate a direct-response fulfillment house that serviced our credit card business. At the time, quite frankly, I thought it was a very unglamorous and potentially boring assignment. But one of the things it taught me is that when you study what appears to be a narrow subject, you've got to broaden the context. That assignment in direct response and direct marketing really gave me an incredible understanding of the power of the direct marketing distribution channel and the importance of customer segmentation. Taking what I learned there, I was able to put together a broader piece of work involving a range of businesses within the company. Suddenly, I was working with people for whom I had no direct management responsibility and figuring out how to get them aligned with a common set of objectives. That was a very important piece of work, both for me and for the company. From a leadership standpoint, it demonstrated to me not just the power of ideas, but also that you do not need formal hierarchical authority to get people to move forward with you alongside common objectives.

- What a great lesson to learn, especially with organizations continuing to flatten. So you are saying that leadership is as much about influencing as it is about anything else?

That is absolutely right. I was an early beneficiary of understanding that the old "command and control" structure was going to be difficult to operate going for ward. My next move at American Express was to the merchandise services division, which at the time was an unprofitable unit that sold a variety of consumer products televisions, luggage, jewelry through the mail to cardmembers. It was not clear how the unit's strategy related to the objectives of the core businesses in particular, the card business. It was, however, a unit very dependent on the direct marketing channel and on understanding customer segmentation. Many people in the company advised me that it would be a serious career mistake to go into that business. But I believed it was important for the company to make an informed decision whether to remain in that business or not; and, on a personal level, I felt that it would be a way for me to gain a wide range of experiences, especially in the marketing and merchandising area. My view was that if the company penalized me for taking a risk, then I probably didn't belong with the company. It turned out to be a terrific opportunity, and I learned a lot of the skills that help me to this day skills such as team building, consensus management, and marketing. Most important, I learned how to take a dispirited organization and galvanize its people to create a turnaround.

- These skills that you picked up along your career certainly helped you lead during and after the terrorist attacks. What are the leadership skills required in crisis situations?

  What is most important is to show compassion and listen very actively while continuing to be a strong and decisive leader. During a crisis, people want direction, but they also need to know that you empathize with them and that you can under stand different reactions and different perspectives. The trust factor is absolutely essential. People can see right through insincerity, and in times of crisis, if you do not engender a level of trust and empathy, there is no way that you can lead your organization through adversity. Particularly in times of crisis, the job of a leader is to redefine reality by defining the new reality. We've got to give our organizations hope, and the hope has to be both factual and inspirational. The other point I would make is that there is no rulebook for leading through a crisis; the way you lead is to follow your values and your beliefs.

- Leaders certainly need to master managing the intangibles as opposed to leveraging tangible assets. What are the "must have" skills for today's CEO, and how have these changed from 5 or 10 years ago?

  Let me name some skills that I believe are constant and then some attributes in which the emphasis may have changed. First and foremost, leaders need courage: courage to make the tough decisions and courage to stick with a strategy that others think will not work. You can assume a certain level of intelligence exists if someone is in the CEO position, but it doesn't always follow that the person possesses courage. Sometimes that doesn't become clear until it's too late. Second, a good leader today needs to be a creative thinker, not just thinking in a step or linear process from a logic standpoint, but thinking in leaps and bounds. Not just connecting the dots but drawing new lines on the page. Third, he or she must have a very high level of integrity. Integrity has always been important, but it is critically important now because we are dealing in an ever-changing environment and a highly uncertain environment. And, increasingly, you're asking an organization's workforce to follow you when you are not providing a hundred percent of the answers. So, they have to believe in you on the basis of your courage, your logic, and the power of your arguments, in combination with trust and faith in your integrity. Now, moving to today's leadership challenges, it is essential to know how to use technology to move a business forward. It's not something that you can ignore any longer. Technology is now so integral to the various business issues and opportunities that we all face, that you have to have a strong understanding of and appreciation for it in order to be fully effective as a leader. It is also critical that you have a global perspective, because resources both human and financial resources as well as raw materials and value-added services are distributed across a wide spectrum, not only geographically but also culturally and ethnically around the world. Without a global perspective, it's going to be difficult to take advantage of the diversity of opportunities in the moderate to long term.

- What is your advice to executives who may be thinking about developing their leadership skills and pursuing a career as a high-level leader?

What I often tell our young executives is to focus on the job at hand. While most people have ambitions, most in fact do not follow up 100 percent on their current commitments. By following up on your commitments I mean execute on the assignment at hand, and place yourself in the shoes of your boss and your peers and your subordinates so that you understand how you need to execute at your current level in order to make those people successful. When you make those people successful, you're going to make the business successful. The second thing I tell up-and-coming executives is that after you have executed on the basic requirements of a job or project, step back and think about what more you can do to add both incremental value and breakthrough value. If people are focused on generating results, adding value, and making the business successful, they can then network off their performance. The third point I try to make is to listen actively to the customer and to ideas being expressed in the organization. You'll get far more insights and information by paying close attention than if you listen in a distracted way.

- What does American Express do in the way of leadership development?

  We look for the key attributes I've mentioned, and we realize that you just don't evaluate those attributes in the annual performance review. You've got to evaluate on a daily basis. A good way to do that is in going through business reviews; ask why a particular result was accomplished or what the actions and behaviors were that led to that result. We look at the quality of thinking and the quality of the execution of each person. We have an environment that encourages people to take risks. You want to create an environment where people have a willingness to stretch and to take some chances, because that's how you really achieve break through results.

- Have you identified positions inside American Express that serve as testing grounds for leadership potential, or do you develop an executive in an existing position that the executive currently holds?

  We have a range of positions that are developmental positions, but we also look at the readiness of the individual. In some cases, we move a person into a job that he or she is fully prepared for in every conceivable way; in other cases, we promote the individual into a job that is a completely new challenge because this is a developmental experience that we want them to go through. So, sometimes we make nontraditional moves because we strongly believe that a particular person will excel in a role despite not being well trained for it. We know that person can handle a broader set of responsibilities, so we create the opportunity within a supportive environment. Our view is that in assessing what moves need to be made, you need to look at the leadership attributes required, the needs of the person, and the needs of the company. We don't have a rigidly structured system that says only these jobs are jobs for developing people. Rather, we are more inclined to match people to the jobs and to the objectives as they change and grow.

- Do you see the future of human capital management in firms as moving toward a 360-degree view of people?
  The so-called war for talent is one of the most important battlefields in business. It is critical that a company create the kind of environment in which people really believe they can learn, grow, and prosper. The focus has to be on developing people.
- American Express has been innovative with its card services. For example, you were the first with the Gold Card, first with the Platinum Card, and now you have the Centurion Card. How have you been able to foster a culture that's built on innovation and built to change, which are important aspects of competitive advantage today?
  What I try to do and what our other leaders try to do is to create both formal and informal forums and a spirit of openness in the company, so that people really feel they have both the opportunity and the obligation to come up with ideas and to innovate. Critical to this is our vision to be the world's most-respected service brand. We therefore have a responsibility and an obligation to constantly increase the value of our products and services to be a superior value provider, which means we can't be a "me too" company; we have to innovate and create. Another way we encourage innovation is to set very lofty objectives, not just in terms of finance but qualitatively as well. The search for qualitative aspirations, frankly, is a journey that we will always be on.

Level Five Leaders

- You recently completed a large-scale five-year study to identify and examine companies that have gone from good to great. Can you give us an overview of the study?

The study grew out of two simple questions: Can a good company or good organization ever become a great one? And, if so, how? When my colleague Jerry Porras and I wrote Built to Last, we left unanswered a gigantic question: If a company doesn't get the right parenting from early on if it doesn't have a David Packard or a George Merck or a Walt Disney to get it off the ground is it basically doomed to mediocrity? General Electric had Charles Coffin right off the bat, so it was always a great company; all its later successes have been based on that early foundation. So, in the latest study, we conducted a systematic research effort to find companies that had made a viable leap from good to great. We began with 1,435 companies that appeared on the *Fortune* 500 list from 1965 to 1995. We looked for companies that showed a sustained performance level no better than market performance in other words, companies whose long-term performance, as compared with the yield on mutual funds, would have grown less than the mutual funds. We looked for companies meeting this criterion; and then we looked at those companies that went through some sort of a leap or a change and afterward, over the next 15 years, enjoyed growth at least threefold greater than mutual funds or the general market. Our findings showed that investments in these companies averaged sevenfold growth. To put this in perspective, GE, over the last 15 years, beat the market 2.8 to 1. We were looking for companies that went from good to doing better than GE at its best, on a dollar-for-dollar-invested basis. The companies we narrowed to, in fact, averaged twice the rate of returns of GE. Using GE as the benchmark of a great company, we found 11 companies that went from good to great, meaning they had better results than GE.

- What is "Level Five" leadership?

Some view leadership as a kind of religion; we didn't want to find a leadership answer to our study in that sense. I used to be a leadership atheist; now I'm a leadership agnostic. It's not that I don't think leadership is important, but in this kind of research, it's not a particularly helpful concept in the sense that it's easy to attribute everything we don't understand or comprehend to leadership. With greater scientific understanding, however, you can begin to understand many more variables at work. So, I really am very, very skeptical of leadership answers. I don't like them. I'm bothered by them. I think we put too much emphasis on them. Infact the evidence suggests that most leaders who are viewed as successful leaders do not produce great companies. Now, that being said, I was very surprised with our findings. The data convinced me that we had a leadership answer. However, we found that it's not leadership itself that takes companies from good to great, because the comparison companies, the companies that tried but failed to make a sustained leap from good to great, had leaders like Lee Iacocca and Stanley Gault and Jack Eckerd. There's no question that these people were real leaders. Even Al Dunlap, some would say, was a leader. But yet, Chrysler, Rubbermaid, Eckerd, and Scott Paper didn't qualify as good-to-great companies. So, because the comparison companies also had leadership, we concluded that the answer to making companies great was not leadership itself. The real distinction was that there is a special form of leadership that takes companies from good to great. What you have in the great companies is Level Five leadership. The comparison companies had Level Four leadership. Here's the difference: Level Five leaders are ambitious first and foremost for the company and its long-term greatness, not for themselves as individuals. As a result, they tend to be personally modest, humble, and reserved, but enormously willful on behalf of the organization. They tend not to become celebrities. They don't have their egos wrapped up in making themselves well known. We identified the only 11 companies in the history of the *Fortune* 500 starting in 1965 that went from good to great using the top benchmarks and standards. In every single one of those cases, the CEOs were largely unknown. They were people like Darwin Smith, Lyle Everingham, David Maxwell, Cork Walgreen, George Cain, and KenIverson. These are not your household name leaders, and yet they absolutely blew away the well-known leaders in terms of actual results. We found that there is an inverse relationship between high-profile, charismatic, egocentric leadership the Level Four leaders and the journey of a company from good to great.

- Are Level Five leaders born or made?
  Some are born, but most are made. That gives the Level Four leaders hope, al though it is hard work getting to a Level Five. Personal growth of any kind is challenging. I believe in the capacity of humans to develop and evolve. That's even one of the great cornerstones of Christianity, the belief that humans can evolve and grow into better people. I believe the same thing with a progression toward Level Five. Most people have the capacity to evolve to Level Five.
- What would be some of the characteristics a person would want to develop to be a Level Five?

Think of it in three layers; the first layer is behavior. Level Five behaviors include things like using "we" instead of "I. " Level Five leaders tend to use "we" and Level Fours tend to use "I. " Another element of Level Five behavior is what we call "the window and the mirror. " Level Fives stand at the window and look for someone or something to attribute success to, even if that success came about through good luck. We asked CEO Alan Wurtzel, "What were the main factors that allowed your company to create this extraordinary result?" He said, "Well, that's easy. The wind was at our backs. " We then showed him the good-to-great stock chart which shows companies' results going along and then his company's results exploding upward.

Meteorologists can pinpoint the landfall of a hurricane within a hundred miles or so several hours before it hits land, making it possible for people to protect their property and flee if necessary.

Earthquakes and volcanoes are the subject of a great deal of research, but scientists are a long way from predicting them with any degree of probability. At best, lessons of preparedness can be learned.

Balancing Policy with Knowledge

You will realize after reading this book that no one is certain that human activities are the cause of the global warming perceived to have taken place during the past 100 years. Indeed, you should wonder if in fact the earth has warmed at all during this time and if the ability to measure such a minute change is possible.

Some scientists and many environmental activists are asking, no, they are demanding, that the governments of the world take drastic action to stop all activities that might be contributing to global warming. Major national and international action plans and treaties have been formulated based on the results of computer climate models. In Chapter 3 we look at the implications of such policies.

Governmental action should be taken only when scientific theories are substantiated more concretely than has been done in the case of global warming. If you read this book with an open mind, you will see that the documented evidence suggests that global warming caused by man's activities is so far unsubstantiated.

Conservation of our natural resources is vital to sustaining human life and society on this planet. Prudent control of our ever-increasing population is a critical task. In the long term, overpopulation, not some imperceptible change in the earth's temperature, will cause the most damage to the human species and its environment.

The earth's temperature has fluctuated throughout geologic time. Species have come and gone. Those species that have adapted are the ones that have survived. Humans have proven to be among the most adaptable of all the species, and we will continue to adapt and survive.

8

Water Resources in Global Climate

Introduction

Water is a principal component of almost every aspect of the global climate, and yet water constitutes the least documented or understood feature of global climate models. Water is ubiquitous and is the only substance to exist naturally in all its phases: solid (ice), liquid (water), and gas (water vapour).

The forms of water affect computer climate models in various and complex ways. Ice and snow are very efficient at reflecting the sun's radiation back into space. Clouds both reflect and absorb the sun's radiation, and they reflect and absorb the earth's radiation as well. On an average day, clouds cover over 60% of the globe, and the cloud cover is ever-changing in location. Water vapour, by contrast, is transparent to most of the sun's radiation but absorbs the earth's radiation, contributing to the greenhouse effect. The oceans absorb the sun's radiation and act as enormous heat sinks that resist temperature change. All these factors greatly complicate the treatment of water in climate models.

Earth's Energy Balance

The complex relationships that all forms of water have with the incoming solar radiation and the outgoing radiation emitted by the earth affect the calculation of the earth's temperature in computer climate models. It is necessary for computer modelers to understand these interactions of the sun's warming radiation with each form of water, as well as their subsequent interactions with the earth's radiation. This simplified picture of energy balance assumes that the incoming radiation from the sun represents 100% of the earth's available energy. We can see that somc 6% is backscattered by the atmosphere, 20% is reflected by clouds, and 4% is reflected by the earth's surface (mainly from ice and snow cover) for a total of 30% reflection loss. In addition, some 50% of the solar radiation is absorbed by the earth. Most of this absorption is by the oceans and lakes, which cover over 70% of the earth, but a significant portion is absorbed by plants. This solar radiation is used in the photosynthesis process that causes plants to grow and is therefore the ultimate source of all our food.

The earth emits radiation at lower energy than the sun because of the earth's much cooler temperature. The sun radiates energy from its surface at 6,000°C (about 11,000°F), and the earth radiates energy from its surface at about 15°C (59°F). The sun's high-energy radiation occurs at short wavelengths (visible and ultraviolet radiation with wavelengths shorter than about 4 micrometers). Because it is so cool, the earth's radiation is at much longer wavelengths (lower infrared energy). The total energy radiated from the earth is minuscule compared with that from the sun.

During the course of the year these numbers would vary depending on the location, season, or even whether it is day or night. The Eskimo will never feel the same solar warmth as the Samoan.

If some of the outgoing energy from the earth is captured by atmospheric gases, and if the incoming energy from the sun is greater than the earth's outgoing radiation, the net result will be a warming of the earth (the greenhouse gas effect). As we saw in Chapter 5, the earth's temperature has fluctuated through the centuries, but only very slowly. The earth's resistance to temperature change is regulated by the water, or hydrological, cycle.

Hydrological Cycle

The global relationships with respect to all water phases and the processes by which one phase changes to another are described by the global hydrological cycle. But even here intricacies are subtle and the consequences are important for the climate. Water evapourates from the oceans and lakes, exists for a time as water vapour, then condenses to form clouds over the earth. The clouds create rain, snow, sleet, or hail, and the water returns to the earth or sea. If the precipitation is on land, some is absorbed by the plants and soils, and the runoff goes into lakes and rivers, eventually returning to the sea. Note that there is more evapouration from the ocean than precipitation into it; so there is a net transfer of water from the oceans to the land. If precipitation were spread evenly over all of the land area of the world, there would be 29 inches of rain per year. Of course, this is not the case; less than ten inches fall each year in the desert areas and over one hundred inches can fall in the rain forests each year.

The water cycle has a major influence on the weather. The sun warms the earth, and the earth's heat is radiated, which in turn causes large masses of warm air to rise into the atmosphere by means of heat convection. In addition, the sun's heat causes water to evapourate from the oceans and lakes. When water vapour cools and condenses to form clouds, it creates a downdraft of cool air. These columns of rising and falling air, coupled with the rotation of the earth and barometric pressure variations, cause the general atmospheric circulation patterns. This circulation is the source of our winds, which are quite different in both direction and speed at the surface of the earth than higher in the atmosphere. The variable circulation creates the trade winds, the jet stream, hurricanes, typhoons, monsoon rains, tornadoes, fog, dust storms, and all other weather events that we either enjoy or must endure.

Table 8.1. The Water Cycle

Process	Flux	
	cm/yr	In/yr
Precipitation on oceans	107	42
Evapouration from oceans	117	46
Precipitation on land	74	29
Evapouration from land	49	19
Runoff from land (river runoff and direct ground water discharge to the oceans)	25	10

This description is an oversimplified version of the atmospheric general circulation model (GCM). Though the general features are easily understood, the patterns are three-dimensional and, because of the many factors that influence them besides the water cycle, not very predictable. The general air circulation of the atmosphere must be modeled very accurately to have any hope of predicting global warming. This is precisely what computer climate modelers strive to do, but they are admittedly a long way from reaching perfection.

We discuss additional complicating factors affecting atmospheric circulation: ocean circulation, ice packs, water vapour, and clouds. It is important to understand the quantities of water found in each of its phases, and Table 8.2 provides an inventory of the water cycle. It is impressive to realize that over 97% of the earth's water is in the oceans, and only 0.001% is in the atmosphere. Only about 2.74% of all the water (rivers and lakes, ice in all forms, and groundwater) on earth is actually in a drinkable form.

Table 8.2. Water/Snow/Ice Inventory Information

Forms of Water	Comments	Volume ($x 10^6 km^3$)	Percentage of the whole
All forms	Total on earth	1408.7	100.0
Sea water		1370.0	97.25
Rivers & Lakes		0.127	0.009
Ground Water	Shallow and Deep	9.5	0.68
Water vapour	Atmospheric	0.013	0.001
Ice	All forms	29.0	2.05

Water Vapour

The atmospheric gas most effective at trapping the earth's radiation is water vapour, which accounts for about 60-70% of the greenhouse effect, according to the International Panel on Climate Change (IPCC). The atmospheric concentration of water vapour varies significantly on a regional basis and with altitude. Overall, however, water vapour is present at much greater concentrations than carbon dioxide or any of the other greenhouse gases. Water vapour generally ranges between 1% and 3% of all the gases in the atmosphere, but it can be as low as 0.1% and as high as 5%, according to Professor S. Manahan.[3] Relative humidity is a measure of the amount of water vapour in the air relative to the maximum amount the air can hold at a given temperature; the higher the temperature, the more water vapour the air can hold. This is the reason that when the relative humidity is 80% or 90% in the deep South, it is really steamy, but in the cool mountains high humidity doesn't seem that wet at all.

The actual humidity at any place and time depends on many factors (cloud cover, barometric pressure, winds, moisture content of soils, etc.) and is not very predictable. The relative humidity can vary from a small percentage to 100% during the course of a day. This short-term regional variation is one of the main reasons why computer climate models cannot predict short-term weather changes, much less longer-term changes.

The issue of water vapour and the lack of knowledge of its role in global computer models prompted David Starr and Harvey Melfi of NASA's Goddard Space Flight Center to organize a workshop on "The Role of Water Vapor in Climate" in 1990. In the foreword of their subsequent report they point out this lack of understanding:... it became clear that present water vapour measurements were inadequate to define the state of the atmosphere on almost any scale; and furthermore, that GCMs general circulation models of the atmosphere lacked the ability to realistically incorporate real moisture data. Our lack of knowledge concerning atmospheric water vapour is all the more striking considering the dominant impact moisture processes have in the Earth's weather and climate.

A goal of the NASA workshop was to assess the scientific state of the knowledge of water vapour in the air and the ability to measure it, and to determine what research should be initiated to improve the situation. They acknowledged the importance of atmospheric water vapour in the following fundamental respects:

1) Water is the principal medium for direct energy exchange among the major components of the Earth System atmosphere, hydrosphere, cryosphere, biosphere and lithosphere.
2) Water vapour is the predominant greenhouse gas and plays a crucial radiative role in the global climate system.
3) Water vapour is an essential ingredient in many atmospheric processes which are intimately involved in determining

specific realizations of climate variations, especially on regional scales.

All these statements were endorsed by the sixty-two eminent climate scientists who represented most of the government research organizations and universities doing state-of-the-art climate model research in the United States. Understanding the role of water vapour in the atmosphere is at the very core of understanding the climate

Clouds

The role of clouds in global climate is even more complex than water vapour alone. Clouds have a significant influence on climate. They can consist of all three water phases ice, liquid, and vapour as well as particulates and trace gases. There are many types of clouds, such as the wispy cirrus clouds high in the troposphere and the cumulonimbus thunderheads formed near the earth. There are many other types of clouds in every part of the troposphere and even some in the stratosphere. Different types of clouds form at different altitudes and under different conditions. They affect the earth beneath in a variety of ways; some cause cooling and some cause warming. Clouds cover over 60% of the earth on any given day, but the effects of each type cannot simply be averaged because of the way the various types interact to either cool or warm the earth. The dynamics of cloud formation and subsequent disappearance are chaotic. The temperature of the earth would be far warmer if there were no clouds; it is generally agreed that clouds cause a cooling effect, and yet climate models often associate an increase in cloud cover with increasing temperature. Let us look at some dominant climate interactions caused by clouds:

Cooling effects:

- Clouds reflect or absorb radiation from the sun (thus the sun's energy never reaches the earth; this affects the land, oceans, and ice and snow cover).
- Clouds cool the surrounding air through evapouration of water droplets (evapourative cooling).
- Clouds transfer water to the earth by means of precipitation (rain, snow, sleet, or hail).

Heating effects:

- Clouds heat the surrounding air by transferring their stored heat (if the temperature of the cloud is warmer than the temperature of the surrounding air like a space heater).
- Clouds reflect or absorb radiation from the earth (thus the atmosphere traps the earth's radiation—the greenhouse effect).
- Clouds radiate energy according to their temperature.

It is easy to see why clouds cause climate modelers fits. If any of the various interactions of clouds are introduced into the models in an exaggerated way, the wrong effect may be emphasized and an erroneous prediction obtained.

As mentioned, the effects of clouds may vary, depending on altitude and type of cloud formation. Another complicating factor is the fact that clouds often form in layers or overlap, another interaction that is difficult to model and that can result in erroneous predictions as well.

Studies of the cloud cover from satellites have shown that, overall, clouds cause cooling, but because cloud dynamics are chaotic and sensitive to turbulent winds, they are still poorly understood. Andrew Revkin, a science writer for Discover, reported on a climatology conference with a focus on clouds. In his article, he quotes Bruce Barkstrom of NASA's Langley Research Center as follows: "Clouds form quickly. They're sensitive to turbulence, which we don't understand at all. We'll be in the next century before we really understand what clouds are doing to Earth's climate."[6] Not only do clouds form quickly and chaotically, they are often much smaller than the resolution of the models. Remember, the area resolution of the models is generally quite large, about the size of New Mexico. When events are smaller than the resolution of the computer model, the model cannot explicitly account for their effects.

It is fair to say that climate modelers have yet to account adequately for the many complex and often opposite cloud effects in their computer models. In early models, scientists simply held the cloud cover as a constant so they would not have to further account for it at all! The early models predicted greater increases in global warming than more recent ones, which are more sophisticated in their cloud treatment. It appears that the more realistic the global climate models become, the smaller the temperature change they predict.

Oceans

If water vapour in the atmosphere and the various cloud effects complicate the understanding of climate, the role of oceans stands as Goliath to David by comparison. Even less information is available with respect to the ocean's circulation. Carl Wunsch of the University of Cambridge states the situation as follows: "Existing knowledge of the ocean circulation is based upon a combination of fragmentary observations (hydrography), and a number of highly plausible theoretical ideas which have rarely been tested directly."7 Whereas many research organizations are making great strides toward a better understanding of ocean circulation and systematic measurement of its important parameters, baseline information is lacking. There have been few long-term programs to obtain the important parameters of temperature, current speed and direction, and salinity, among others, relative to depth in the oceans. Not only is the oceanic data base fragmentary, there is very little of it.

The fact that the oceans absorb about half of the solar energy that reaches the earth is essential to the behavior of the global climate. This energy is not evenly distributed, most of it being absorbed within the area between 30° N and 30° S latitudes. The oceans generally heat up slowly during the summers and release heat to the atmosphere as they cool down during the winters; thus, they are gigantic heat sinks that moderate the earth's temperature. In addition, some of the warm water near the equator moves toward the poles and provides one of the sources for ocean currents. Note that oceans and continents do not receive an equal distribution of radiation in the Northern and Southern Hemispheres.

- It may mix with deeper, colder ocean water, but only during the winter.
- It may mix with deeper, colder ocean water all year.
- It may never mix with deeper, colder ocean water.

This mixing also helps to drive ocean currents. Like the atmosphere, the ocean has a circulation, driven by the movement of the huge masses of warmer and cooler water. Cold water is more dense than warm water. Pure water is most dense when its temperature is 4°C (39°F). The cold water sinks toward the bottom of the ocean, and warm water rises toward the surface. These movements help to create the major ocean currents. The currents are further affected by the topography of the ocean floor, which influences areas of upwelling and down welling. Ocean circulation is further complicated by the saltiness, or salinity, of the water, which affects its density and temperature. There are certain salinity and temperature ranges that create layers in the ocean. (If you have ever gone swimming in a lake, you have undoubtedly felt a colder layer of water a few feet below the surface.) These layers are so pronounced in the oceans that oceanographers give them names. Pearn Niiler of Scripps Institution of Oceanography at La Jolla, California, describes five layers in the Atlantic Ocean, emanating from distinct sources, as follows:

- Antarctic Bottom Water—The most dense and deepest is below 3500 meters, with a source on the Antarctic Continental Shelf.
- North Atlantic Deep Water—lies between 4000 meters and 2000 meters and appears as emanating from north of the Gulf Stream. It can be traced to the Norwegian and Labrador Seas.
- Mediterranean Water—Another salt tongue at the 1500 meter level that can be traced eastward to the Mediterranean Sea.
- Antarctic Intermediate Water—emanates from the Circumpolar Current area at 800-1000 meters.
- Sub-Tropical Mode Water—the most salty is a lens on the surface caused ostensibly by high evapouration under the Trade Winds.

This is the major reason that the climate modelers must make do with "hokey" ocean models, as described by Stephen Schneider, when commenting on the ocean model used by Jim Hansen. (Hansen, you will recall, is the scientist who testified before Congress that "global warming had begun.")

There are, very broadly, two types of ocean current flow, laminar and turbulent. Laminar flow is smooth and orderly, like a deep running river, flowing slowly and peacefully. It can be described by means of simple mathematical equations. By contrast, turbulent flow is like a churning rapid in a narrow gorge. Turbulent flow is nonlinear and therefore chaotic, with whirlpools and eddies, and it is characterized by a great variety of flow speeds. By its very nature this nonlinear, chaotic process is nearly impossible to model mathematically. Unfortunately, there is a great deal of turbulent flow within ocean currents. Scientists David Neelin, of the University of California at Los Angeles, and Jochem Marotzke, of Massachusetts Institute of Technology, in their article in Science, "Representing Ocean Eddies in Climate Models," call the mathematical treatment of ocean eddies one of the main challenges in modeling the ocean. They point out that ocean eddies are small in scale when compared to the grid size used in ocean models, which means that they must be included in the models by indirect mathematical expressions rather than explicit, or fundamental, equations.[9] This has not been done very well in the past. The problem is analogous to the atmospheric problem of clouds, in which clouds are often small with respect to the size of the model grid.

Processes in the oceans occur at a much slower pace than on land and in the atmosphere. Water vapour lasts only about ten days in the atmosphere before it becomes part of a cloud or condenses back into liquid. Water in the soils lasts maybe a year. By contrast, the residence time of water in the oceans is roughly 3,000 years.[10] Given these tremendous time differences, you can see how difficult it must be to couple ocean models with atmospheric climate. There are ocean effects that are very slow and cause climate effects over decades. In early models these ocean effects were ignored by climate modelers or greatly oversimplified. Only recently have computer modeling experiments been reasonably successful using realistic ocean models. The computer time required to incorporate a good ocean model into an atmospheric circulation model is enormous. The ocean models require at least as much computer time as atmospheric general circulation models, and when coupled together the resulting time requirements multiply rather than add. This could add weeks to the time required for a single experiment of a century-long simulation.

Cryosphere (Ice and Snow)

The ice sheets of Antarctica and Greenland, the mountain glaciers, and winter snow cover are very important in global climate models. The earth's albedo is the total reflection of the sun's radiation back into space. The average albedo for the earth and its atmosphere is 30%. However, different substances vary markedly in albedo. Table 8.3 gives the albedo values for various types of surfaces. Ice, snow, and white clouds can reflect 70-80% of the sun's radiation, whereas oceans and land reflect very little, only 2-6%, absorbing most of the solar radiation. The variation in albedo numbers reflects the fact that sunlight reflection depends on the angle of the sun with respect to the earth's surface, or the angle of incidence.

Important statistics about the water-ice inventory are presented in Table 8.2 and Table 8.4. Of particular interest is the overwhelming amount of water that rests in the oceans, the almost trivial amount in the atmosphere, and the great variation of snow and ice coverage over both the land and sea from winter maxima to summer minima.

The significant variations in the amount of snow and ice from season to season are illustrated in Table 8.4. At the peak of winter snow coverage in the Northern Hemisphere, about 50% of the land surface is covered and 10% of the sea surface is covered. In the summer only about 2% of the land and 5% of the ocean are covered.

Table 8.3. Albedo Values for Various Earth Surfaces

Surface	Albedo (%)
Ice or fresh snow	75-95
Old snow	40-60
Sand or deserts	18-35
Grasslands or green crops	15-25
Forests	14-20
Dense or evergreen forests	7-15
Cities	14-18
Plowed fields	10-15
Asphalt	about 8
Water (lakes and oceans)	2-6
Clouds (depending on type)	20-70

Remember that it is summer in the Southern Hemisphere when it is winter in the Northern Hemisphere and vice versa. These major seasonal variations must be accurately included in the computer models for accurate results. The oceans, including the ice-ocean atmosphere interface, pose major obstacles to the validity of global computer models.

An additional important factor in the water cycle is the extent and depth of the permafrost. Change in the permafrost is measured in centuries rather than decades. The coverage of the permafrost regulates the amount of water evapourated into the atmosphere over great portions of the higher latitudes (near the poles) in both hemispheres.

Table 8.4. Winter and Summer % Snow/Ice Coverage

	Northern hemisphere		Southern hemisphere	
	land	ocean	land	ocean
Summer	2.0%	5.2%	28.6%	1.2%
Winter	49.0%	9.7%	> 28.6%	9.7%
	Land + Ocean		Land + Ocean	
Summer	4%		6.5%	
Winter	25%		13.5%	

The ice sheets are also important in that they contain about 80% of the fresh water on earth. If the ice sheets start to melt, they will add fresh water to the oceans, which will affect the salinity of the oceans around the ice sheets. This imbalance in salinity will cause changes in the ocean-ice-atmosphere relationship, which will in turn create long-term (hundreds of years) changes in ocean circulation.

Predicted Ocean Rise with Global Warming

There have been dire predictions of a large sea level rise associated with the doubling of carbon dioxide in the popular myth of global warming. Though these predictions are mostly exaggerations, the IPCC report summarized estimates from different researchers of upward to 3.65 meters (12 feet). However, the IPCC report proposed more modest values of from 9-29 centimeters (3.5-11.4 inches) for probable ocean rise due to global warming.

Table 8.5. Predicted Contribution to Sea Level Rise (1985-2030)

Source	Low estimate (cm)	High estimate (cm)	Best estimate (cm)
Thermal expansion	6.8	14.9	10.1
Mountain glaciers	2.3	10.3	7.0
Greenland ice sheet	0.5	3.7	1.8

Antarctica ice sheet	-0.8	0.0	-0.6
Total	8.7	28.9	18.3

If oceans are going to rise significantly because of global warming, which was a major theme of environmentalists at the end of the 1980s, the bulk of the increase would have to be due to breakup and melting of the ice sheets of Antarctica and Greenland. These ice sheets are much more important to sea level rise than the mountain glaciers in northern Europe and America; if all mountain glaciers were to melt, the sea level rise would only be about 25-45 centimeters (10-18 inches), according to the IPCC. The consensus of the scientists in the IPCC is that the ice sheets will not break up or melt in the near future; scientific assessment of potential ice sheet melting indicates that this would take centuries to happen.

Recent studies of the variation in ice and snow accumulations over the Greenland ice sheet for the past 18,000 years, carried out by Professor W. R. Kapsner and a research group from Pennsylvania State University and the University of Washington, have found that variation in storm tracks, not variation in temperature, is the important factor for determining precipitation variability. In discussing these results, David Bromwich, of Byrd Polar Research Center at Ohio State University, states:

The realization that precipitation variability over the polar ice sheets depends on far more than just temperature changes means that predicting the effects of "global warming" on sea level is not as straightforward as attempted in the report by the Intergovernmental Panel on Climate Change.

Bromwich goes on to discuss the need for better spatial resolution in GCMs before they can adequately assess these storm track versus temperature effects all of which indicates yet another area in which computer climate models do not include an adequate mathematical model of the climate.

EL NINO

El Nino is a major climate event that occurs every three to five years; it affects climate all over the world, causing both flooding and drought. It has no known relationship to greenhouse gases or global warming, but it has profound global effects. If computer models are to be used for predicting climate 100 years in the future, they must first be able to predict successfully such major oceanic events as El Nino.

El Nino is a natural, quasi-periodic change in the sea-surface temperature of the ocean currents off the coast of Ecuador and northern Peru in South America. Sea-surface temperature can be abnormally warm or cool. If it is warm, it is called El Nino, and if cool, La Nina. El Nino generally begins near Christmas and hence the name, which is Spanish for little boy, or "the Christ Child."

During El Nino, there is a major warming of the ocean surface temperature near South America, with an associated change in atmospheric pressures and trade winds. The pressure change is a "seesaw" effect between the southeastern tropical Pacific and the Australian-Indonesian region. When one of these areas has abnormally high pressure, the other has abnormally low pressure, and vice versa. The pressure phenomenon was observed independently and termed the Southern Oscillation. Since this connection between the Southern Oscillation and El Nino has been recognized, the acronym ENSO (El Nino-Southern Oscillation) has been used, according to Henry Diaz and Vera Markgraf in their recent monograph El Nino, Historical and Paleoclimatic Aspects of the Southern Oscillation.

It is now accepted that this major and natural quasi-periodic event affects the global climate in a real and significant way. There are areas of increased drought and other areas of increased rainfall at far-reaching places on the globe. The drought-wet conditions in Southern California, the southeastern United States, and central Africa all seem to have connection to ENSO. For example, a group of scientists from Columbia University and Zimbabwe recently observed a relationship between the maize yield in Zimbabwe and ENSO.

Scientists do not fully understand the causes of El Nino. There are computer models that have been somewhat successful in predicting El Nino events. However, according to Richard Kerr, writer for Science, the computer models have been inconsistent, for example, missing the prolonged El Nino beginning in 1993. The GCMs used to predict global warming are essentially the same computer models used to predict El Nino, but they have not incorporated the El Nino components when predicting global warming due to increasing greenhouse gases.

All aspects of the global water cycle discussed in this chapter water vapour, clouds, the oceans, and ice constitute the most important components of computer climate models. All processes in the ocean are affected by the amount of cloud cover, the diurnal variation in the sun's radiation, the earth's rotation, the pull of the moon's gravity, which causes the tides, and the seasonal effect of the earth's orbit around the sun. It seems clear that these hydrological effects are the least well known features of the computer models.

9

Flora and Fauna in locating Position

Introduction

Global warming could destroy or fundamentally alter a third of the world's plant and animal habitats within a century, bringing extinction to thousands of species, according to a study by Great Britain's World Wide Fund for Nature, an affiliate of the World Wildlife Fund. The study said that the most vulnerable plant and animal species will be in Arctic and mountain areas, where as many as 20 percent could be driven to extinction. In the north of Canada, Russia and Scandinavia, where warming was predicted to be most rapid, up to 70 percent of habitat could be lost, according to this study. The report, Global Warming and Terrestrial Bio-diversity Decline, was written by Jay Malcolm, professor of forestry at Toronto University, and Adam Markham, the executive director of Clean Air/ Cool Planet. Pests and weedy species would fare best, they said, concluding, "If past fastest rates of migration are a good proxy for what can be attained in a warming world, then radical reductions in greenhouse gas emissions are urgently required to reduce the threat of bio-diversity loss". To adapt and survive the expected rate of warming during the next century, the report said that plants may need to move 10 times more quickly than they did when recolonizing previously glaciated land at the end of the last ice age. Few plant species can move at a rate of one kilometer per year, the speed that will be required in many parts of the world.

Dangers of Tropical Deforestation

The Earth had about 5.2 million square miles of old-growth forest remaining by about 1995. An estimated deforestation rate of 62,000 square miles a year, the amount being logged in the late 1990s, could reduce that figure to zero within two human lifetimes.

More than 50 percent of the world's forests have been destroyed within the past 100 years. An area equivalent to 57 soccer fields falls each minute. Trees are roughly 50 percent carbon. Alive, they remove carbon dioxide from the air and replace it with oxygen. Dying, dead, or being burned as fuel, trees become sources, not absorbers ("sinks") of carbon dioxide and other greenhouse gases. The destruction of forests around the world is no trivial matter for the atmosphere's carbon budget. While the atmosphere contains about 750 billion tons of carbon dioxide, forests contain about 2,000 billion tons. Roughly 500 billion tons is stored in trees and shrubs and 1,500 billion tons in peat bogs, soil, and forest litter.

Stephen Schwartzman, a senior scientist with the International Program of the Environmental Defense Fund, sketches the scope of tropical deforestation:

An area of forest bigger than Belgium, Holland and Austria put together, or about 40 percent of California, was cut down and burned every year between 1980 and 1995, some 62,000 square miles per year. NASA's Landsat satellite photographs show that more than 200,000 square miles, an area about the size of France, has been cleared and burned in Brazil alone. All of this has happened since the 1970s.

Deforestation increased rapidly during the 1980s. Three million acres of forest were being lost per year in Indonesia, providing 70 percent of the world's plywood and 40 percent of its tropical hardwoods. By the early 1980s in the tropical forests of Indonesia, roughly 1.2 percent of remaining coverage was being felled each year. In 1988, 48,000 square miles of Amazon rain forest were burned to clear land for farming. Nearly overnight, most of the burned land was turned from a net producer of oxygen to a source of carbon dioxide and methane.

As Indonesia's forests were being felled, government policy encouraged a homesteading program on Sumatra and Borneo for farmers moving from overcrowded Java. Indonesia was using some of the same government incentives Brazil was using to "develop" the Amazon Valley. During the 1980s, hundreds of thousands of Brazilians moved to the Amazon Valley in a fashion resembling the homesteading of the North American West a century earlier. Some of this migration was financed by international development agencies, such as the World Bank.

In Indonesia and Brazil, the new landowners began altering the landscape by felling portions of the rain forest, then burning what they had cut. The rainforest canopy, which had heretofore produced a surplus of oxygen, was replaced by cattle, human beings, and motor vehicles. A survey conducted by the United Nations in 1982 estimated that tropical forests were being felled at a rate of 70,000 square miles (180,000 square kilometers) per year. The report estimated that 300 million subsistence farmers around the world, most of them in the tropics, were turning an area of forest the size of Denmark into a net producer of carbon every three months. In 1987, 25,000 square miles of Brazilian rainforest were felled; a year later, the area that was deforested doubled. In the 1980s, lowland tropical forests in Indonesia, the Philippines, Malaysia, and parts of West Africa shrank quickly as well.

According to the Brazilian government's own figures, the amount of Amazonian forest lost to fire doubled between 1997 and 1998. The Woods Hole Research Center, in Massachusetts, said that 7,800 square miles of Amazonian rain forest caught fire in 1997. About 80 percent of the fires were set on purpose, according to the United Nations; 42 million acres of rain forest (an area the size of Florida) were being lost per year by the 1990s.

Given deforestation and other land-use changes in the Amazon Valley, the area no longer functions generally as "the lungs of the world." The area acts as a carbon sink (absorber) only when temperature and precipitation patterns are healthy for the forest. In years of unusual drought (such as during the El Nino episodes of the middle and late 1990s), unusually hot, dry weather turns the Amazon into a net producer of carbon dioxide. In a usual year, the Amazon absorbs about 700 million tons of carbon dioxide, but in the recent El Nino year of 1995, the Amazon forest added 200 million tons of carbon dioxide to the atmosphere. According to A. Lindroth, A. Grelle, and A.S. Moren, writing in Nature, a decrease in soil moisture can turn a carbon dioxide consuming forest into one that adds the gas to the atmosphere; this variation occurs in temperate zone forests as well as those of the tropics, the authors assert.

By the 1990s, according to R.A. Houghton and colleagues, "This large area of tropical forest is nearly balanced with respect to carbon".

The combined effects of deforestation, abandonment, logging, and fire may thus yield sources of carbon. These fluxes are similar in magnitude (but opposite in sign) to the sink calculated recently for natural ecosystems in the region. Taken together, the sources (from land-use change and fire) and the sinks (in natural forests) suggest that the net flux of carbon between Brazilian Amazonia and the atmosphere may be nearly zero, on average.

Threats to the last remnants of a nation's forests sometimes provoke action. According to Stephen Schwartzman, "China, not a world leader in green consciousness, last year 1998 banned all logging in its few remaining natural forests after disastrous flooding wreaked havoc along heavily populated rivers. In so doing, China hoped to save remnants of forest cover on the upper headwaters. But so much forest is already gone that it may not make much difference".

Deforestation has taken the trees from the world's tropical forests to some notable venues. Vintage ships of the British Royal Navy, for example, were refurbished during the 1990s with large amounts of mahogany harvested illegally from an indigenous reserve in the Amazon River basin. The harvest raised a ruckus among English environmentalists, including Friends of the Earth, as the defense ministry's mahogany purchase was questioned on national television. More than 7,000 cubic feet of mahogany was cut, moving by ten to twelve trucks a day from the Kayapo indigenous reserve in Para state to the town of Redencao, home of the Juary logging company, harvester of the lumber. Each log was harvested illegally under Brazilian law. Despite the illegality of the harvest (aided by slack enforcement) the mahogany was purchased for the British ministry of defense by an English supplier, Parker Kislingbury.

In February 1999, deforestation was implicated as a major culprit when dozens of people died and hundreds of millions of dollars in property was destroyed in massive floods which shut down the industrial capital of South America, Sao Paulo. The same year flooding, also aggravated by deforestation, provoked unprecedented destruction in Caracas, Venezuela. Deforestation also intensified the death and destruction wrought in Central America by Hurricane Mitch.

About half of the rain which falls on a rainforest is produced by the forest's own humidity. Deforestation over large areas influences regional climate because there are fewer plants to produce the moisture that returns to the Earth as rain. More water runs off, carrying more topsoil, leaving less forest cover. Deforestation thus breaks a forest's hydrology cycle. As a tropical forest declines, regional weather becomes hotter, drier, and more prone to fire within surviving stands of trees. Increasing wildfires (such as those during the late 1990s in Mexico, Brazil, and Indonesia) add even more carbon to the atmosphere. By the late 1990s, the burning of tropical forests was contributing about 20 percent of the human-induced carbon dioxide buildup in the atmosphere. The burning of the Amazon rainforest alone contributes about a quarter of the worldwide total. The "lungs" of the world are being turned, within living memory, into yet another anthropogenic source of greenhouse gases.

According to Schwartzman,

The Woods Hole Research Center has found that for every acre cleared and burned in the Amazon, at least another acre burns in ground fires under the forest canopy or is degraded by selective logging (not picked up by the satellites). The frequency and extent of these ground fires skyrocket in El Nino events, which can then cause drought in some tropical forests. Such fires are likely to increase in frequency and intensity with global warming.

Deforestation is especially dangerous because once an area is denuded, local soil conditions (and sometimes weather) changes in such a way as to make the regrowth of trees more difficult. In many areas where deforestation has been extensive, previously forested areas have been replaced by sparse grasses, stunted shrubs, and bare, eroded earth.

"The scale of the problem is mind-boggling," said Daniel C. Nepstad of the Woods Hole Research Center, who has been surveying deforestation in the Amazon Valley. Nepstad estimated that 400,000 square kilometers of rainforest, an area twenty times that of Massachusetts, became vulnerable to fire during 1998, compared to an average of 15,000 square kilometers a year which had been cleared and burned during the previous few years. "Once burned," wrote Nepstad, "Amazonian forests become more susceptible to future burning. Wildlife is killed or dispersed (except for some species that thrive on disturbance). Timber, forest medicines and forest fruits are destroyed, and a large part of the carbon contained in the trees is gradually released into the atmosphere".

A major problem, according to Nepstad, "is that the rainforest is available in abundance and is therefore very cheap. When forest is cheap and labor and capital are scarce, it is the forest itself that becomes the fertilizer, the pesticide, the herbicide, the plow". Nepstad suggested "[a] new model of rural development in Amazonia, which restricts access to most (about 70 percent) of the region's forests, and which creates a situation of land scarcity, higher land prices, and higher investments in agricultural production systems". Nepstad believes that people will value the forest only when it becomes an economic good in human financial terms. People will quit burning the forests when they can earn more money by sustaining it. About half of the area burned each year in Amazonia is accidental, according to Nepstad, who recommends prohibition of burning at times when fires may easily spread out of control.

Destruction of the Amazon rainforest accelerated toward the end of the twentieth century. According to studies by Nepstad, the forest was being destroyed two to three times more quickly than previous estimates. Nepstad conducted his research with scientists at the Institute of Environmental Research in Belem, Brazil. They measured forest losses at 1,104 sample points from a light plane and on the ground. They also interviewed more than 1,500 mill operators and landholders in the Amazon region. Nepstad and his associates assert that satellite imagery, on which most previous estimates have been based, "often fails to distinguish between pristine forest and burned or cut forest land newly covered with fast-growing brush".

Nepstad said that roughly 17,000 square miles of Amazon forest were lost to cutting and burning in 1998, about three times the official Brazilian government estimate of 5,700 square miles. Nepstad also said that the total amount of rainforest already lost to human encroachment is about 217,000 square miles, or about 16 percent of the entire Amazon Valley. Nepstad cautioned that these figures are conservative and that "the real numbers could be significantly larger".

Research published during October 1998 by the Hadley Center, part of the United Kingdom Meteorological Office, anticipates that large areas of "closed" tropical forests will die as the world warms during the next century. Instead of soaking up pollution, these forests will dump more carbon dioxide into the atmosphere than all the world's power stations and cars have produced during the past 30 years. "Closed" tropical forests, according to the Hadley Center report, contain trees covering a large proportion of the ground where grass does not form a continuous layer on the forest floor. Forests of this type declined from 289,700 thousand hectares (1980) in Africa to 241,500 in 1990. In Latin America, tropical forests declined from 825,000 to 753,000 hectares during the same period, and in Asia, tropical forest coverage fell from 334,500 hectares in 1980 to 287,500 in 1990.

Deforestation sometimes places additional debits on the world account of greenhouse gases. When tropical forests are cut and slashed, for example, the resulting piles of dead wood attract termites, whose populations explode as they feast on a food supply that has suddenly increased. Termites very efficiently digest carbon, turning it into methane.

At the November 1998 conference on climate change in Buenos Aires, Britain's Hadley Center for Climate Prediction and Research presented a report indicating that large parts of the Amazon Valley could become desiccated by 2050. It was said that this desiccation may threaten the world with an unstoppable greenhouse effect. The Hadley report asserts that land temperatures will rise by an average 10 degrees F. over the next century. In addition to intensifying aridity in parts of tropical South America, large parts of tropical Africa are expected to become desiccated by 2050, according to the Hadley Center researchers. These climate changes will force roughly 30 million people to face intense hunger, the report asserts.

By about the year 2050, according to one study, "a decrease in annual rainfall of up to 500 millimeters in some key areas, combined with temperature rises of up to seven degrees C., will begin to kill tropical forests, creating deserts". At this point, the study says that these forests will become carbon sources rather than absorbers (sinks), as their remains burn or rot. "Up to 2050, the land surface takes up carbon through an increased growing season, which is being measured at the moment. But when we move to 2050, the tropical dieback releases so much carbon to the atmosphere it causes the concentrations to increase. This may enhance the build-up of carbon dioxide in the atmosphere".

Hurricane Mitch and Deforestation

As early as 1984, economist Ian Cherrrett, working with a Dutch nongovernmental organization in Honduras, witnessed "total destruction of the environment going on at an accelerated rate," and warned that if forest cover was not protected, "as far as I'm concerned, within 20 years it's going to be too late." Orin Langelle of the Vermont-based Action for Community and Ecology in the Rainforests of Central America (ACERCA) observed, "Fourteen years later, Hurricane Mitch confirmed his prediction".

Five years before flooding from Hurricane Mitch devastated Honduras, J. Almendares, a Honduran medical doctor, warned readers of the British medical journal Lancet that deforestation was making the country more vulnerable than ever to deadly flooding. Almendares presented evidence that "desiccation and soil erosion caused by cattle grazing and sugarcane and cotton cultivation have altered the regional hydrological cycle". These changes have led to fewer rainy days, but more intense downpours.

Almandares presented temperature statistics from one deforested area of Honduras indicating that the average ambient air temperature had risen 7.5 degrees C. between 1972 and 1990. In neighboring Nicaragua, during the final months of the Sandinista revolution, similar temperature rises were reported in Managua after dictator Anastasio Somoza ordered the wholesale destruction of trees in the city to deny Sandinistas places to hide during gun battles.

Central America had 200,000 square miles of forest in 1900. By the 1980s, only 36,000 square miles survived, and the rate of deforestation was increasing, especially in the Miskito region of Nicaragua and Honduras. A 1998 report by United Nations agencies and nongovernmental organizations documented a regional deforestation rate of 958,360 acres (1,500 square miles) a year.

Honduras lost a third of its forests between 1964 and 1990. Honduran forests continued to be felled at a rate of 80,000 hectares a year during the 1990s, a rate which, if sustained, amounts to a quarter of remaining forested land per decade. In the meantime, people who can no longer wrest a living from denuded (or corporate-controlled) land have been moving to Honduran cities, where malaria has become endemic. "Blood transfusions have now become a significant means of malaria transmission in Honduras," Almendares and Sierra wrote.

During October, 1998, Hurricane Mitch made landfall on Central America's Miskito Coast with winds as strong as 178 miles an hour, dropping as much as three feet of rain. Crossing the mountains, Mitch turned northwest, through El Salvador, Guatemala, and southern Mexico. By the time the storm reached the sea again, it had killed 10,000 people and left nearly three million others homeless. In Honduras, thousands of people who survived the storm lost their jobs in devastated banana plantations.

The devastation wrought by Mitch raised questions in Central America not only about potential increases in hurricane severity due to global warming, but also about changes in land use across the region which makes many areas more prone to severe flooding during heavy rains. The same questions were raised in Caracas, Venezuela, after devastating floods occurred there late in 1999.

Bill Weinburg described the conditions which have made hurricane flooding so devastating in Central America:

Fire and flood fuel each other in a vicious cycle. Landless peasants colonize the agricultural frontier, or clear forested slopes for their milpas (fields). The more forest is destroyed, the more the hydrologic cycle is disrupted; with no canopy for transpiration, local rainfall and cloud cover decline; aridity makes the surviving forest vulnerable to wildfires. Then, when the rains do come, sweeping in from the Caribbean on the trade winds, there are no roots to hold the soil and absorb the water. Millennias' accumulated wealth of organic matter is swept from the mountainsides in deluges of mud. Tlaloc, the revered Nahua rain god who the Maya called Chac Mool, brings destruction instead of abundance.

In Nicaragua, 2,000 people died in the Chinandega municipality of Posoltega, as ten communities were buried in mudslides when the Casita volcano crater collapsed. Three-quarters of a million people were left homeless in the area. Posoltega lies in an area that has been almost completely deforested. Hurricane Mitch's trail of death and damage in eastern Nicaragua also was intensified by deforestation of the upper Rio Coco watershed. More than 20 inches of rain caused the river to rise more than 60 feet within a few days.

Preservation of forests became an environmental issue in Eastern Nicaragua following Hurricane Mitch's devastation. Local native peoples and ecologists united that same year to evict Korean-owned timber giant Solcarsa, which had won a government contract in a large area of the Miskito rainforest. According to Weinburg, at least 370,500 acres were being deforested annually in Nicaragua at the time of Hurricane Mitch. Nicaragua has lost 60 percent of its forest cover within the last two generations.

Hurricane Mitch left Nicaragua with an enduring legacy, as described by Weinburg:

In September 1999, almost a year after the disaster. President Aleman declared a national emergency over a plague of rats in Nicaragua. Their numbers exploded from an overabundance of dead meat both human and animal following the hurricane. The rats overran fields and homes, decimating crops. Poison had to be distributed in mass quantities to beat the infestation.

Former Honduran President Rafael Callejas attributed the extreme death and damage wrought by Hurricane Mitch to "mudslides that were the result of uncontrolled deforestation and therefore could have been prevented". Ecological activism can be as dangerous in Honduras as the winds, rains, and mudslides of a major hurricane. A few weeks before Mitch made landfall, Carlos Luna, a local opponent of logging in the central mountains near Tegucigalpa, the country's capital and largest city, was gunned down by unknown assailants.

Deforestation has become a major problem (and political issue) in Mexico as well as in Central America. During the spring of 1999, fires raged out of control throughout the mountains of southern Mexico, sweeping through parts of the Chipas highlands, as well as the Sierra Tarahumara of Chihuahua. The fires cloaked Mexican skies in an acrid haze from its borders with Guatemala and Texas. Among the causes of these fires are private timber operations on ejido (Native-held) lands that compound the deforestation caused by campesinos clearing lands for their milpas. In October 1997, deforestation took a deadly toll when Hurricane Paulina hit the Sierra Madre del Sur in the Mexican states of Guerrero and Oaxaca. Mudslides cascading down denuded mountainsides left scores dead and thousands homeless, especially in Zapotec country.

The main cause of tropical deforestation perhaps two-thirds of it is slash-and-burn agriculture. Much of the deforestation is caused, in turn, by small farmers and ranchers who are threatened by the spread of commercial farming and ranching enterprises. Driven out of more populous areas, subsistence farmers are forced to destroy large amounts of tropical forest to create new farmland. According to a study by the Global Futures Bulletin, the pattern is the same in much of Africa, Asia and Latin America. Increasing human population and pressure on the land is the ultimate cause of the deforestation of slash-and-burn agriculture, according to the report.

Fire, Carbon and Boreal Forests

A report by Greenpeace International suggests that between 50 and 90 percent of the Earth's existing boreal forests are likely to disappear if atmospheric levels of carbon dioxide and other greenhouse gases double. These forests comprise a third of the Earth's remaining tree cover, about 15 million square kilometers, across Russia (where they are called "taiga"), Canada, the United States, Scandinavia, and parts of the Korean Peninsula, China, Mongolia, and Japan. Large forests also clothe many mountain ranges outside of these zones. In total, boreal forests cover about 10 percent of the world's land area.

The Greenpeace report indicates that global warming 's toll on the boreal forests had begun by the early 1990s. The report warns that decaying forests may provide an extra boost to rising carbon dioxide levels, causing warming to feed upon itself. The decline of boreal forests also endangers more than one million indigenous people who live in them, including the Dene and Cree of Canada, the Sami (Lapplanders) of Norway, Sweden, and Finland, the Ainu of northern Japan, and the Nenets, Yakut, Udege, and Altaisk of Siberia. Some of the forests' larger animals, such as the Siberian tiger, already are near extinction. The Green peace report concludes that rapid logging of the boreal forests is intensifying pressure on animal life, and accelerating the release of even more carbon dioxide and other greenhouse gases into the atmosphere.

Climate change could imperil New England's maple syrup industry, as a gradual rise in global temperatures strips New England of its native sugar maples. A report by the U.S. Office of Science and Technology Policy said that expected climate changes within the present century are likely to shift the ideal range for some North American forest species northward by as much as 300 miles, exceeding the species' ability to migrate naturally. If emissions of greenhouse gases continue to increase, the report projects that maples will recede from all areas of the continental United States except the northern tip of Maine.

The Sierra Club also has described the possible effects of global warming on sugar maples:

Regions dependent on forests for various commodities, such as the maple-syrup producing states of New England and the Midwest, could also face devastating losses. EPA projects that by 2050 the range for the sugar maple could shift north of all but the northernmost tip of New England. This possibility has serious implications for the maple syrup industry, which currently provides up to $40 million annually to the regional economy.

The same Sierra Club report traces increasing temperatures and insect infestations in boreal forests to a rise in temperatures which began in 1976.

With rising temperatures have come larger and more frequent forest fires, according to Green peace. "Unless atmospheric concentrations of greenhouse gases are quickly stabilized," the report states, "climate-vegetation models predict that large areas of boreal forest will be reduced to patchy open woodland and grassland, resulting in lowered biological diversity and a reduced ability to store carbon".

If boreal forests continue to decline, their burning and rotting could contribute to the release of as much as 225 billion tons of extra carbon dioxide into the atmosphere, raising current levels by a third, accelerating the pace of warming. Trees could have difficulty colonizing the thawing tundra to the north of their present ranges because they simply cannot migrate quickly enough, and because the treeless tundra cannot evolve quickly enough to sustain them.

Kevin Jardine sketches the role of insect outbreaks in the anticipated destruction of boreal forests:

Pests which may invade boreal forests under warming conditions include the western spruce budworm, the Douglas-fir tussock moth, and the mountain pine beetle. Kurz [and colleagues] conclude that "biospheric feedbacks from temperate and boreal forest ecosystems will be positive feedbacks that further enhance the carbon content of the global atmosphere". By 1998, spruce budworms had devoured 50 million acres (20 million hectares) of Alaskan woodland.

In moderation, the forest usually benefits from insect outbreaks because they reduce the likelihood of catastrophic fire by helping to eliminate older stands and diseased trees. However, populations of insects such as bark beetles, the Siberian silkworm and the spruce budworm can explode and devastate millions of acres of forest. The life cycles of the spruce budworm (Choristoneura fumiferana) and the spruce bark beetle (Ips typographus) are strongly influenced by climate, with both species likely to increase in numbers during the kind of warmer, drier weather predicted in a global warming world. A Canadian government study released in 1987 showed that conifers were growing on average up to 65 percent slower than in the 1940s and 1950s, because of spruce budworm outbreaks, and possibly acid rain.

William K. Stevens of the New York Times described unprecedented destruction of boreal forests in Alaska by spruce bark beetles that have been whipped into a reproductive frenzy by a warming environment:

Once these purple-gray stretches of tree skeletons were a green and vital part of the spruce-larch-aspen tapestry that makes up the taiga. Today, in a stretch of 300 or 400 miles reaching westward from the Richardson Highway, north of Valdez, past Anchorage, and down through the Kenai Peninsula, armies of spruce bark beetles are destroying the spruce canopy or have already done so. Often the trees are red instead of gray freshly killed but not yet desiccated.

On the south-central coast of Alaska, cool temperatures have heretofore kept the spruce bark beetle under control. As temperatures have warmed, however, the beetles have killed much of the tree cover across three million acres, one of the largest insect-caused forest devastations in North America's history.

While warming increases many insects' reproductive energies, warming destroys many trees' reproductive capacities. According to Jardine, rising temperatures can cause boreal trees' pollen and seed cones to develop too rapidly because of higher-than-usual spring temperatures, leading to reproductive failure. Boreal seeds also germinate within a specific range of soil temperatures. For example, writes Jardine, black spruce seeds germinate between 15 and 28 degrees C. "If the soil temperature falls below 15 degrees," writes Jardine, "the processes that cause germination come to a halt. If soil temperatures rise above 28 degrees C., bacteria and fungi can attack and consume seeds. The probability of germination also declines rapidly for higher temperatures".

Valerie A. Barber, Glen Patrick Juday, and Bruce P. Finney studied tree-ring data from Alaskan forests for 90 years during the twentieth century, and found that warmer temperatures inhibit the growth of Alaskan white spruce, thereby decreasing boreal forests' ability to draw carbon dioxide from the atmosphere.

The tree-ring records show a strong and consistent relationship over the last 90 years, and indicate that, in contrast with earlier predictions, radial growth has decreased with increasing temperature. Our data show that temperature-induced drought stress has disproportionally affected the most rapidly growing white spruce. If this limitation in growth due to drought stress is sustained, the future capacity of the northern latitudes to sequester carbon may be less than currently expected.

R. Suffling correlates increases in temperature to a rise in the amount of land lost to forest fires in the boreal forests of Scandinavia and Canada. Given the unplanned nature of forest fires, policy responses will be limited, Suffling warns, and he speculates whether increasing pressure to harvest remaining forests for human use will accelerate this process. He describes a wave of forest fires in the boreal forests of Canada: "Since the mid-1970s, there has been a massive fire outbreak in response to a series of warm, dry summers".

Work by W.R. Emanuel and colleagues indicates that boreal forests on Earth may become nearly extinct within roughly a century if the world warms to the extent forecast by many climate models.

During the summer of 1995, large fires scorched forests in parts of northern and central Canada, consuming as many as 240,000 acres a day. A study by the Canadian Forest Service concluded, "The northern forest has lost almost a fifth of its biomass over the last twenty years because of enormous increases in fires and insect outbreaks". In 1970, these forests were absorbing 118 million tons of carbon dioxide each year, but during the 1990s, the same area had become a net producer of 57 million tons of carbon dioxide per year.

Canada's boreal forests may be reduced to a fraction of their current range by warming temperatures, according to a governmental report. The area covered by boreal forests in Canada declined 20 percent between 1975 and 1995. The report, compiled by Environment Canada in 1994, predicted an average winter warming on the coast of British Columbia of four degrees C. in the winter and 2.5 degrees C in summer by 2050. Such a change could raise snow levels as much as 3,000 feet and reduce mountain snowpack to between one-half and one-sixth its size in 1995, according to the report. If, over the next century, there is a doubling of atmospheric carbon dioxide and a warming of roughly 9 to 10 degrees F., yellow birch, sugar maple, hemlock, spruce, beech, and other trees which now flourish in the temperate zones would be forced to move northward.

W.D. Billings describes some of the problems which confront plants because of the speed of anthropogenic warming:

If the predicted rates of climatic warming during the twenty-first century hold in the Arctic, the change may be far more rapid than any climatic changes during deglaciation, or even during the Little Ice Age. Will this warming be much faster than the ability of plant migration and soil formation to keep up? Perhaps. For tree taxa in the Great Lakes states, future range extensions will have to occur at least ten times as fast as the average Holocene rate, or 200 kilometers percentury, to track the expected temperature rise.

The flora of the future will favor plants best adapted to increased carbon dioxide levels as well as a warmer environment. Swift changes in temperatures also will favor plants which can adapt to weather extremes and those which can move quickly. These attributes describe most "weedy shrubs and grasses which directly compete with young trees in unforested areas. New forests will not be established easily, and the world will become much weedier". In addition, mid-latitude forests which perish because they cannot beat the heat to higher latitudes will release their own carbon when they die.

According to George M. Woodwell, director of the Woods Hole Research Institute, a one degree C. change in temperature is equivalent in the mid-latitudes to between 60 and 100 miles of latitudinal change. Woodwell sketches the speed of the climate changes suggested by many climate models: "With warming in the range of tenths of a degree per decade and the highest warming rates in the higher latitudes, the changes in climatic zones will be of the order of miles per year".

Insect infestations also have been increasing in forests that are accustomed to sustained heat, as well as in cooler boreal forests. By 1998, the southeastern United States was suffering through its worst infestation of armyworms in recent memory, as the pests chewed through thousands of acres of crops, pasture, and turf. The armyworm population explosion was caused by a mild winter following a summer drought, according to Richard Sprenkel, pest management specialist with the University of Florida's Institute of Food and Agricultural Sciences.

Armyworms balloon from the width of two hairs to the thickness of a pencil during a life span of three to four weeks. A field of crops can be wiped clean by the insects within 48 hours. Armyworms have been eating corn, cotton, peanuts, and grasses used for pasturing animals. Armyworms also threaten lawns, golf courses, and athletic fields. The insects do most of their damage during a frenzy which consumes the last two days of their lives. Insecticides do little good at this stage, since the armyworms have already laid the eggs which will hatch into a new generation. The rise of the armyworms has been linked to the increasing rarity of hard winter freezes in the Southeast.

Animal Adaptations to Warming

The migration and breeding patterns of many animals were responding to warmer temperatures. For example, Jerram Brown has charted the breeding seasons of Mexican jays in the Chiricahua Mountains of southern Arizona for 31 years. By 1998, Brown found that the jays were laying their eggs an average of 10 days earlier than they did in 1971. Camille Parmesan has analyzed records tracking the distribution patterns of 57 nonmigratory butterfly species across Europe. She found that, during the last century, two-thirds of the species have shifted their ranges northward, some of them by as much as 240 kilometers. "We ruled out all other obvious factors, such as habitat change," said Parmesan. "The only factor that correlated was climate".

Parmesan and collegues' tracking of butterfly ranges provided evidence of poleward shifts in entire species' ranges. In a sample of 35 non migratory European butterflies, 63 percent have ranges that shifted to the north by 35 to 240 kilometers during the twentieth century, while only three percent of ranges have shifted to the south. The study team's evaluation of its data ends with a warning about other species:

Given the relatively slight warming in this century compared with anticipated temperature increases of 2.1 to 4.6 degrees C. for the next century, our data indicate that future climate warming could become a major force in shifting species distributions. But it remains to be seen how many species will be able to extend their northern range margins substantially across the highly fragmented landscapes of northern Europe. This could prove difficult for all but the most efficient colonizers.

Warmer temperatures were bringing butterflies to England two weeks to a month earlier than during the 1970s. Scientists studying 35 of the estimated 60 species of British butterflies say that some, such as the red admiral, can now be seen a month earlier. Others, "such as the peacock and the orange tip, are appearing between 15 and 25 days earlier than two decades ago," according to a report in the Times of London. Roy and Tim Sparks, both of the Centre for Ecology and Hydrology at Monks Wood, Cambridge shire, analyzed data from 1976 to 1998 provided by the Butterfly Monitoring Scheme, whose members check more than 100 sites each week from April to September. According to a news report in The Guardian of London, "One [red admiral] was monitored after crossing the [English] Channel on New Year's Day".

Some birds, like butterflies, have been extending their ranges northward in Europe since 1970. Chris D. Thomas and Jack J. Lennon of the University of Leeds' School of Biology analyzed the breeding distribution for birds in Britain, finding that their northern ranges had expanded northward an average of 18.9 kilometers between roughly 1970 and 1990. The authors note that temperatures warmed in Britain during this period, "which we propose might be casually related".

Global Warming and Fisheries

Drastic declines in some western Alaskan salmon populations during 1997 and 1998 led some observers to ask whether rising water temperatures had played a role, because salmon are very sensitive to temperature. According to a report by the World Wildlife Fund,

While salmon can withstand higher temperatures in summer when food is abundant, in the winter their tolerance drops considerably. As cold-blooded creatures, their metabolism increases in warmer water and keeping up with this high metabolism requires large amounts of food. If sufficient food is not available, salmon can starve.

The decline of Pacific salmon runs during the late 1990s suggests that global climate change could devastate fish populations on which millions of people rely for food. In Alaska during 1997 and 1998, few salmon returned from the ocean to spawn at their birthplaces. Those which did return were smaller than average and arrived later than usual. The key factor in the decline of salmon runs appears to have been the fact that water temperatures in 1997 and 1998 were much higher than usual.

For Canada's western regions, climate models forecast an increase in precipitation, water runoff, and flooding in winter and a decrease in precipitation and runoff during summer. Higher winter river flows are expected to damage salmon spawning grounds, reduce survival and growth of fish because of increased stream temperatures, and damage Fraser River salmon due to increased predation by warm-water species. In 1995, the Canadian Department of Fisheries and Oceans blamed a collapse of Fraser River salmon runs on predation by mackerel, which invaded the salmon spawning grounds along with warmer-than-average ocean waters provoked by El Nino conditions. At the same time (and according to the Canadian government, for the same reasons) the Queen Charlotte chinook salmon runs declined about 80 percent.

The salmon catch in Scotland was 35 percent less in 1999 than during 1998, which itself was the second worst year since comprehensive records began being kept in 1952. The reduced catch, due in part to warming water temperatures, is threatening the livelihood of river proprietors, country hotels and bed and breakfasts, as well as ghillies (guides) who help the angler find his fish. Warmer sea water decreases the population of krill which the salmon eat. Anglers also say that sea lice from salmon reared on fish farms are infesting wild salmon, killing them. The fish farms also produce a polluting slurry. In 8 of 32 rivers on the west coast of Scotland, salmon were virtually extinct by the year 2000. Seals are killing salmon in some rivers as well.

Cod (a cold-water fish similar in some ways to salmon) also has been declining as habitats warm. In mid-July, 2000, the World Wildlife Fund (WWF) placed North Sea Cod, staple of British fish and chips (as well as a $60 million annual fishery), on its endangered-species list. Populations of North Sea Cod have declined 90 percent in 30 years, according to the WWF. The cod have been over-fished, and their population is falling because they do not breed well in warmer water. The North Sea in the year 2000 is as much as three degrees C. warmer than it was in 1970.

Global warming could result in an estimated eight percent decrease in fish yields worldwide. Global warming also may significantly damage recreational fishing. An Environmental Protection Agency (EPA) report issued in 1995 shows that several cool-water fish species, especially trout, would diminish as waters warm. This study projects that between eight and ten states (of the United States of America) could lose all of their cool-water sports fisheries within 50 to 60 years. Between 11 and 16 other states could lose half of their cool-water sports fisheries, according to the study.

Rising temperatures in British Columbia's largest sockeye salmon spawning river, the Fraser, provoked a government shutdown of commercial salmon fishing at the height of the season late in September 1998. According to an Environmental News Service dispatch from Victoria, British Columbia, above-average temperatures in the river impaired the salmon's swimming and jumping abilities, and afflicted them with "proliferating pathogens and an invasion of warm-water predators such as squaw fish". The article also described fears that "seawater temperatures 1.5 degrees C. above average in the adjacent Georgia Strait could trigger toxic algae blooms".

Between a quarter and two-thirds of salmon returning to various locations in the Fraser river system were dying before they could spawn, many of them from conditions related to rapid warming of their aquatic environment. The water warmed because of the below-average snowpack and above-normal temperatures in the interior of British Columbia, the Fraser's drainage basin. A large aluminum smelter also was delivering large amounts of heated water to an upstream tributary of the Fraser, adding yet another human provocation to the salmon's warming environment. Clear-cutting of upstream forests is also blamed for some of the warming of the river's waters.

Above-average water temperatures in this area also had caused salmon to die in unusually large numbers during 1992 and 1994. If the temperature rises two degrees more (from a late-summer peak of 21 degrees C. to 23 degrees C.) none of the salmon will survive the swim upstream to spawn. At 23 degrees C., most salmon will die of heat prostration. The warming of the river and resulting fishing shutdown in September, 1998, brought out angry fishermen, who threatened to block cruise ships in Vancouver's busy harbor.

Robert Gough comments,

Global (primarily marine) fisheries provide the world with about 20 percent of its animal protein. Worldwide fisheries are pressed to their natural limits and international fishing fleets are at capacity. Fishing fleets of various nations exceed their own boundaries and encroach on the territorial waters of other countries.

According to the Intergovernmental Panel on Climate Change (IPCC), global warming may cause some fisheries to collapse, while others expand. As Gough comments,

Climate changes can exacerbate the effects of over-fishing at a time of inherent instability in world fisheries. In addition, over-fishing creates an increased imbalance in the age composition of a stock, and may reduce the resiliency of the population. Further, changes in ocean currents may result in changes in fish population location and abundance and the loss of certain fish populations.

Fish populations in many areas also probably will be affected by interruptions in their breeding cycles caused by saltwater intrusion into estuaries as seas rise. Roughly 70 percent of the world's fish use shoreline waters in their breeding cycles. Writes Gough, "Fish production will thus suffer when such nursery habitats are lost". Gough warns that many species in the sea and on the land which have adapted to specific environments may not be able to adapt to the speed of coming temperature changes: "Surviving species may succumb to predatory pressure and competition from more exotic species better adapted to the new conditions".

Global Warming and Agriculture

Many people who do not farm for a living share a stereotype of agriculture as a family affair, a builder of character, and a style of employment which evokes, for tillers of the soil, a basic sense of enjoyment from communion with nature. During the twentieth century, however, agriculture has become progressively more mechanized on a massive scale suited to large, worldwide markets. Agriculture, like other modes of production in our machine culture, has come to demand less human labor and an increasing amount of fossil-fuel energy. Industrial-scale agriculture also requires copious amounts of synthetic fertilizers. For all except a few remaining (and often struggling) family farmers, agriculture has become as industrialized as factory work.

Ecologist Barry Commoner described how the American farm has changed: Between 1950 and 1970, the total U.S. crop output increased by 38 percent, although the acreage decreased by 4 percent and the labor [number of people employed] fell by 58 percent. This sharp increase in productivity was accomplished by an 18 percent increase in the use of machinery and a 295 percent increase in the application of synthetic pesticides and fertilizer.

How will agriculture fare in a warmer world? The MINK [Missouri-Iowa Nebraska-Kansas] Study surveyed potential climate change in the central United States, North America's agricultural heartland. Under certain circumstances, the authors found, higher levels of carbon dioxide might enhance growth of some crops, but as a whole, "under the best of these scenarios the productivity of the region's agriculture would be significantly diminished". Agriculture would be severely affected not only by heat stress, but also by reduced surface-water supplies, since most global climate models predict that as the atmosphere warms the interiors of continents would become not only hotter, but also drier, especially during the growing season. An additional problem facing farmers in Nebraska and Kansas is depletion and salinization of aquifers which already support a large part of agricultural production in both states, especially their drier western areas.

Craig Benjamin, writing in Native Americas, describes how large scale mono-cultural farms make themselves vulnerable to a rising risk of pest attack in a warmer, more humid world.

This impressive vulnerability of industrial agriculture is key to understanding how climate change will likely have an impact on global agriculture and on the relationship between industrial agriculture and indigenous farming communities. Faced with rapid and dramatic climate change, the impressively vulnerable industrial farm can conceivably continue to use large-scale irrigation and artificial fertilizers to counter the effects of changing temperature and precipitation.

While some skeptics argue that global warming will benefit agriculture by providing plants with a higher level of carbon dioxide, Pim Martens contends that increased growth may be counterbalanced by molds and other parasites which thrive best in hot, humid weather. "It is generally believed that a climate change will have negative effects for global food production," Martens writes. To cite one of many examples, the Mediterranean Fruit Fly could expand into Northern Europe during the next century with the degree of global warming projected by the IPCC.

Research by Fakhri A. Bazzaz and Eric D. Fajer casts doubt on the skeptics' assertions that a carbon dioxide enriched atmosphere will lead to more plant growth and greater agricultural yields.

Studies have shown that an isolated case of a plant's positive response to increased CO2 levels does not necessarily translate into increased growth for entire plant communities. Photosynthetic rates are not always greatest in CO2-enriched environments. Often plants growing under such conditions initially show increased photosynthesis, but over time this rate falls and approaches that of plants growing under today's carbon-dioxide levels. When nutrient, water, or light levels are low, many plants show only a slight CO2 fertilization effect.

Bazzaz and Fajer assert, "We do not expect that agricultural yields will necessarily improve in a CO2-rich future". William R. Cline adds that scarcity of water (which is forecast for many continental interiors as the atmosphere warms) also may reduce agricultural yields.

Studies by Martin Parry estimate that the European corn borer could move 165 to 200 kilometers northward (in the northern hemisphere) with each one-degree C. rise in temperature. The potato leaf-hopper, a major pest for soybeans, presently spends its winters along the Gulf Coast. Global warming could move this range northward. The range of the hornfly, which caused about $700 million in damage to beef and diary cattle across the United States during the late 1980s, could be similarly affected.

N.C. Bhattacharya has found that while enriched carbon dioxide causes accelerated growth in most plants, others respond negatively. Increasing the growth rate of plants also tends to accelerate depletion of soils, stunting later growth. Heat also may be detrimental to some plants even as their growth is being stimulated by rising carbon dioxide levels in the atmosphere.

Y.A. Izrael summarizes the rough road of agriculture in a warmer world:

Estimates of the impact of doubled CO2 on crop potential have shown that in the northern mid-latitudes summer droughts will reduce potential production by 10 to 30 percent. The impact of climate change on agriculture in all, or most, food-exporting regions will entail an average cost of world agricultural production [of] no less than 10 percent.

Martin Parry elaborates, "While global levels of food production can probably be maintained in the face of climate change, the cost of this could be substantial". Gains in production at higher latitudes are unlikely to balance reductions in the hotter mid-latitudes, which are major grain exporters today.

Harold W. Bernard, in Global Warming: Signs to Watch For, describes global warming's anticipated role in the spread of wheat rust, a fungus which thrives in dry heat. Wheat rust destroyed millions of tons of wheat in North America during the Dust Bowl decade of the 1930s. At the same time, the pale western cutworm, another pest which favors hot and dry conditions, damaged thousands of acres of wheat in Canada and Montana.

Cline models global warming three centuries into the future. By that time, he expects that agricultural production in many contemporary breadbaskets will have been devastated. Indicating an average July maximum for Iowa between 100 and 108 degrees F., Cline scoffs at the idea that higher carbon dioxide levels in the atmosphere may enhance agricultural yields. By the year 2275, Cline indicates that temperatures may become so hot that most staple grain crops in present day Iowa (and surrounding states) may die of heat stress. Wheat, barley, oats, and rye simply will not grow in the climate which Cline projects for the United States Midwest roughly three centuries from today. In Eurasia, rice, corn, and sorghum will be pressed to their limits by such temperatures as far north as Moscow. By 2275, according to Cline, the atmosphere's carbon dioxide load may be eight times the levels of the 1990s.

In the much shorter range (at twice contemporary levels of greenhouse gases, a level expected by many climate models by the end of the twenty-first century), the United States is forecast to experience a decline of perhaps 20 percent in agricultural production, while some areas of Russia and northern Europe may see production increases. A beneficiary of moderate global warming could be Iceland, which, according to Cline, is expected to see its agricultural production rise 50 percent. For comparative purposes, Cline remarks that the average temperature in the United States during the 1930s Dust Bowl years was only 0.9 degrees C. higher than the 1951–1980 average range. After the twenty-first century, following Cline's analysis (which assumes increasing use of fossil fuels), farmers will be dealing with a degree of warmth that probably will be outside human historical experience.

According to Thomas R. Karl and associates, increases in minimum temperatures are important because of their effect on agriculture. Observations over land areas during the latter half of the twentieth century indicate that average minimum temperatures have increased at a rate more than 50 percent greater than that of maximums.

The rise in minimum temperatures has lengthened the growing (frost-free) season in many parts of the United States; in the Northeast, for example, Karl and associates write that the frost-free season began an average of 11 days earlier during the 1990s than during the 1950s. The compression of daily high and low temperatures may be related to increasing cloud cover and evaporative cooling in many areas, Karl and associates propose. Clouds depress daytime temperatures because they reflect sunlight, as they also warm night-time temperatures by inhibiting loss of heat from the surface. "Greater amounts of moisture in the soil from additional precipitation and cloudiness inhibit daytime temperature increases because part of the solar energy goes into evaporating this moisture," they write.

In a warmer world, erosion and drought will probably pose greater dangers to agriculture. This is not the paradox that it seems, because, according to Karl and associates "Not only will a warmer world be likely to have more precipitation, but the average precipitation event is likely to be heavier". Karl cites data indicating that heavy precipitation events increased roughly 25 percent from the beginning to the end of the twentieth century. Karl describes how a generally wetter world also will be a place in which drought may threaten flora and fauna more often:

As incredible as it may seem with all this precipitation, the soil in North America, southern Europe, and in several other places is actually expected to become drier in the coming decades. Dry soil is of particular concern because of its far-reaching effects, for instance, on crop yields, groundwater resources, lake and river ecosystems. Several models now project significant increases in the severity of drought.

Karl and associates temper this statement by citing studies indicating that increased cloud cover may reduce evaporation in some of the areas which are expected to become drier. All in all, however, most of the world's flora and fauna will suffer markedly in a significantly warmer world.

10

Policy Implications and Responses

Introduction

The hysteria generated over the global warming myth has led to calls for drastic governmental action. Many of the demands are extreme, with the potential for adverse economic and environmental consequences. The issue has inspired debate among scientists, economists, politicians, and environmental activists: What should the actions be, considering the many uncertainties that exist in the scientific understanding of the causes and effects, as well as the extent and direction, of climate change? Should any action be taken at all? Who should decide?

Identifying the Problem

Let us first define the problem that government is being asked to solve. The generally accepted view of global warming holds that increasing greenhouse gases in the atmosphere, especially carbon dioxide, will lead to higher global temperatures, adversely affecting the earth's climate and thus its biodiversity, societal patterns, and land use. According to the theory, human activities contribute to most of this greenhouse gas increase, and therefore major changes must be made in these activities to reverse the trend of increasing carbon dioxide. What these changes should be and whether or not they should be made are at the root of governmental policy decisions.

Anthropogenic, or human-caused, carbon dioxide emissions come from two principal sources: burning of fossil fuels (coal, oil, gas) and changes in land-use patterns (deforestation, loss of farmland). Later we consider these sources in detail and discuss their relative effects on the atmosphere, as well as the costs of reducing emissions or continuing on the path of "business as usual."

The global warming theory has been developed by climate scientists, based on years of observation and research and on the recently developed computer climate models. From scientific research to government policy, especially international government policy, there is a large gap. That gap has been bridged by the environmental movement, with its activists, by the scientists who support the global warming theory, by scientific institutions that benefit from research funding, and by politicians.

The environmental movement had its first real influence on government policy when Silent Spring, Rachel Carson's 1962 book about pesticides, raised fears and prompted action by the government to ban DDT and other pesticides. Other books, such as Paul Ehrlich 's The Population Bomb, warned of the dangers of overpopulation and its effect on diminishing food supplies and the fragile environment. These books raised the public's consciousness about population growth and environmental degradation, spawning environmental organizations and stimulating efforts by many people to reduce waste and conserve energy. However, the books also contributed to the fears that had been raised and fed the hysteria of overreaction. Later publications from Ehrlich and others built on these fears and escalated calls for quick, often Draconian, solutions.

In the nineties the public and political response to environmental issues is more sharply pronounced. The threat of global warming caused by anthropogenic greenhouse gases has put carbon dioxide emissions in the spotlight. There are essentially three stances taken with regard to these emissions. One is that no action should be taken to cut emissions because there is no problem. Another is that, before taking drastic action, research should continue on the possibilities of climate change, the consequences of such change, and the appropriate steps to take to affect the change or adapt to it. The last position is that, regardless of the uncertainties in our present knowledge, we should begin to take drastic action "just in case," as a sort of insurance policy.

While scientific studies, government reports, and policy statements continue to emphasize the uncertainties that exist in the computer climate models and thus in the global warming issue, the most popular approach is to proceed as if the threat is real. For example, in its 1992 report, "Policy Implications of Global Warming", the National Academy of Sciences, a scientific society that serves as advisor to the federal government on scientific and technical matters, concluded that even with the "considerable uncertainties," global warming "poses a potential threat sufficient to merit prompt response." This position reinforces the demands of environmental activists as well as the scientists such as Stephen Schneider and James Hansen, who have invested major effort in the study of climate change.

The governmental actions that are being considered could have profound effects on the economies of many nations, and for that reason, many economists, as well as many scientists, urge caution and more study before taking drastic action. It is important that policymakers consider how much effect limiting or reducing carbon dioxide emissions will have on the atmosphere and at what economic and social costs.

Brief History of Greenhouse Politics

Before discussing the various proposed actions and their potential consequences, let's see how government became involved in the global warming debate. Although the greenhouse effect was described in the 1820s, and even associated with carbon dioxide in the 1860s, it was well over a century before the issue was noticed outside the scientific community.

During the 1960s an increase in environmental research, aided by the development of mainframe computers, stirred up some political interest. Conferences were held and reports were written, but no recommendations for action were made, and other, more compelling issues held the public's attention, issues ranging from Vietnam to the race for space, from the population explosion and civil rights to "sex, drugs, and rock and roll."

In the seventies, disillusioned with big business and buoyed by campus demonstrations, the spirit of Woodstock, and a liberated lifestyle, the "new generation" swept the country; the time was ripe for awareness and activism. Environmental issues became immensely popular, and global cooling caught the imagination of hippies and politicians alike it was the decade of one of the coldest winters on record. It was also during this decade that scientists became worried about deterioration of the ozone layer, a concern that led to the ban of chlorofluorocarbons (CFCs) in aerosol cans.

The presidential administrations of the seventies, primarily those of Nixon and Carter, were committed to environmental action. Under Richard Nixon, the Clean Air Act of 1970 was enacted, giving extensive power to the federal government. Earlier attempts to cope with dramatic increases in air pollution had left most of the jurisdiction to the states, with the federal government acting as technical advisor and financial assistant. With little progress being made, Congress stepped in with the Clean Air Act, which created a whole set of standards and compliance requirements. The U.S. Environmental Protection Agency (EPA) was created at this time and given the authority to set the standards. It remains the most powerful environmental agency in the country and perhaps even in the world.

Under Jimmy Carter there was extensive support of alternative energy, paving the way for major research and development on renewable energy technologies, especially solar energy. The demand for energy had grown rapidly in the 1960s and early 1970s. Energy was inexpensive and little thought was given to efficiency. However, according to a 1993 American Scientist article, the "oil shocks" of the early and late 1970s, when OPEC (the Organization of Petroleum Exporting Countries) raised oil prices, led to concern and changes in the way Americans used energy. Public concern was high in 1977 when Carter declared that energy use was the most important issue facing the country. Fuel conservation and efficiency became popular, leading to a further slowdown in the demand for energy. Ironically, this slowdown alleviated much of the concern over air pollution and lack of fuel supplies, and it set the stage for the 1980s, when energy prices would drop once again and environmental problems would take a back seat to more pressing issues.

The decade of the seventies ended with the First World Climate Conference in Geneva, Switzerland. By this time the climate predictions had shifted from cool to warm, and this 1979 conference covered environmental issues such as the greenhouse effect and climate change, prompting further research and a heightened interest among scientists and governmental agencies.

The following year, the United Nations Environment Programme (UNEP), the World Meteorological Organization (WMO), and the International Council of Scientific Unions (ICSU) sponsored a meeting in Villach, Austria. The attendees concluded that carbon dioxide-induced climate change was a major environmental issue and agreed that more research was needed for a firm scientific base. These organizations pointed out the need for global cooperation between developing and industrialized nations.

In that same year, Ronald Reagan was elected president of the United States, and there was a decided shift in environmental philosophy in the government. Reagan came under fire from environmentalists because he was a friend of business and industry, and, in fact, it was under the Reagan administration that the budget for the EPA was slashed dramatically and federal investment in solar energy and other renewable energy sources virtually ended. In addition, the demand for energy had dropped, and energy prices as well, so that the public political climate favored Reagan's position.

In 1983 the EPA issued a report warning of a 2°C rise in global temperature and determined that the only way to avoid it was to ban the use of coal before the year 2000. But at the same time the National Academy of Sciences (NAS) issued a report downplaying the problem, saying that they did "not believe ... that the evidence at hand about CO_2-induced climate change would support steps to change current fuel-use patterns away from fossil fuels." In addition, the NAS report recommended

> caution in undertaking any major changes in current behavior and policies solely on account of CO_2. It is probably wiser not to act aggressively right now when we really do not know the future consequences or context of CO_2 increase.

In spite of this inconsistency between the positions of the EPA and the NAS, the idea of an environmental calamity quickly caught the imagination of the news media. In 1985, a second conference was held in Villach, Austria, sponsored once again by UNEP, WMO, and ICSU, and attended by research scientists, ecologists, and computer modelers. According to Sonja Boehmer-Christiansen's 1994 commentary in Nature, this symposium attracted support from major environment/energy research bodies but was attended by only two government scientists, one from the United Kingdom and one from the United States (Department of Energy). Thus the primary interests being served were the contract research institutions with interests in carbon dioxide and climate variability. Boehmer-Christiansen points out that, with such a strong warning about the dangers of global warming, "stringent regulations would be needed opening up energy markets to 'green' technologies."[7] The warnings spread, and the political momentum needed for major governmental action was rapidly building.

Scientific meetings, press coverage, confirmation of the ozone "hole," and a fully engaged environmental movement resulted in the 1987 Montreal Protocol, the first international treaty based on the results of computer models! The main purpose of the treaty was to eliminate CFCs entirely because of their implication in the diminishing stratospheric ozone. Shortly thereafter, the greenhouse effect of CFCs was recognized, and the Montreal Protocol was amended to hasten the phase-out period of these chemicals.

The year 1988 stands out as a watershed year in the story of global warming policy. That year's extremely hot summer, along with droughts in the United States and flooding in Bangladesh, provided the ideal setting for the global warming debate to shift dramatically to the side of the alarmists.

In June an international conference in Toronto, "The Changing Atmosphere: Implications for Global Security," called for reduction of carbon dioxide emissions by approximately 20% of 1988 levels by the year 2005. The conference proposed a world fund, supported by a carbon tax, to help developing nations cope with the special economic problems they faced in complying with emissions reductions. In addition, the Intergovernmental Panel on Climate Change (IPCC) was created by WMO and UNEP to coordinate worldwide climate research efforts. This organization is now the foremost international authority on climate change. The reports published by the panel are used as a major source of information for policymaking.

In 1988 George Bush was elected president after campaigning as the "environmental candidate." However, after Bush took office, he continued Reagan's cautious approach to environmental policy and found himself criticized by environmental organizations and by other countries.

Hansen's testimony was controversial in more ways than one. When the Office of Management and Budget (OMB), which monitors federal policy statements, insisted the following year on attaching a qualifying statement to his written testimony, cries of censorship were heard. Then-SenatorAl Gore considered the OMB's act to be part of Bush's effort to downplay environmental issues: "Why would the Bush White House go to such lengths to avoid facing the facts about the environment? Is it because the necessary changes would cause some political risk?" And what was the OMB's caveat that caused such a commotion? This is the statement that was added: " these changes should be viewed as estimates from evolving computer models and not as reliable predictions." In fact, this statement is consistent with the qualifying statements that appear in essentially all scientific reports on climate change research.

The international call for action became louder as the European Economic Community (EEC) criticized America for inaction, and in 1989 the European Council of Environment Ministers called for an immediate response to the global warming crisis regardless of uncertainties. The official White House position, which was to wait and see before taking actions that might be inappropriate further down the road, was influenced by a report from the George C. Marshall Institute, according to David E. Newton of the University of San Francisco, in his reference handbook Global Warming. The Marshall Institute provides scientific advice for public policymakers, and its conclusions, criticized by the global warming alarmists, support a conservative view of potential climate change. Ironically, at the very time the United States was being criticized for inaction, Congress was passing its amended Clean Air Act, which would cost $20 billion per year.

At international conferences in 1989 and 1990, President Bush blocked proposals for specific limitations of carbon dioxide emissions because of his concern for the economic impacts of drastic reductions.

His actions upset environmentalists worldwide and led to further criticism of his administration.

In 1990 IPCC published its first report, Climate Change, which supported the view that emissions resulting from human activities were increasing atmospheric greenhouse gas concentrations and would contribute to a warming of the earth's surface. While stopping short of making actual recommendations for action, the report stated that an immediate reduction of 60% in anthropogenic emissions would be required to stabilize the concentrations at 1990 levels. The ambiguity of the report, however, led to a wide range of interpretations. The global warming adherents pointed to the warnings in the report while detractors quoted the inconsistencies.

As heat records continued to pile up and Nature reported a decrease in the Arctic ice cap, the international debate intensified. The 1992 United Nations Conference on Environment and Development (UNCED) in Rio de Janeiro, a conference better known as the Earth Summit, called for drastic reductions in carbon dioxide emissions. President Bush further alienated himself from environmental activists when he announced that he would not sign the treaty unless specifics about the levels at which carbon dioxide would be stabilized were eliminated. Concessions were made, leaving out specific emissions levels and time frames, but the United States was criticized for watering down the agreement.

The Framework Convention for Climate Change that was agreed upon at Rio was formulated by the IPCC. The prime goal of the agreement was to stabilize atmospheric greenhouse gases in order to prevent "dangerous" anthropogenic interference with the climate system. However, according to Bert Bolin, IPCC's chairman, the agreement contained language and concepts that should be clarified, such as the term "dangerous." So while recommending action, no specifics were given, leading to a great deal of flexibility in interpreting the responsibilities of the various governments.

In 1993 a new "environmental president" took office, along with an "environmental vice president." Although Vice President Al Gore still takes an extreme view of the dangers of global warming, as evidenced by his pronouncement on Earth Day 1994, the Clinton administration is more cautious than environmental activists would like. These activists have criticized the administration's Climate

Action Plan of 1993 for favoring industry. The plan relies mostly on voluntary programs and increased use of renewable energy sources, and it calls for "limited, and focused, government action and innovative public/private partnerships." A spokesperson for Greenpeace, one of the more aggressive environmental organizations, complained that "all we're getting is 'greenwash' without any real change in energy policy," while another one said that the "very industries called upon to be partners in the plan believe climate change is a myth." While the Clinton plan continues to be lambasted by the activists, it is buying time for further research and consideration of the consequences of carbon dioxide emissions reductions.

Charting Today's Course

With so many organizations some scientific, some environmental, some political involved in trying to influence policy responses, you can be sure of a wide variety of recommendations. The IPCC's scientific assessment is used as a basis for much of the policy being made, but, as pointed out in the 1990 report, it is a summary of scientific understanding at the time of the report, not a definitive statement about climate change.

Uncertainties attach to almost every aspect of the issue, yet policymakers are looking for clear guidance from scientists; hence authors have been asked to provide their best-estimates wherever possible [IPCC emphasis], together with an assessment of the uncertainties.

The IPCC members clearly recognize the perilous position they are in. They have been criticized both for being too conservative and for overreacting.

The 1992 IPCC report includes some updates that would affect much of the policy response being considered. Of particular note is its reevaluation of Global Warming Potential (GWP), a rating that represents the relative potential climate effect of greenhouse gases. The GWP is used as a tool in determining which gases should be considered when calling for emissions reductions. (It was used to rate CFC replacements, and some, which were otherwise acceptable, were eliminated because of the GWP, leaving very few products that can be used as substitutes.) The 1992 IPCC report reveals increased uncertainty in the GWP calculations.16 If these calculations cannot be trusted by scientists, how can they be used as a basis for determining the amount of reduction needed in carbon dioxide emissions? If scientists cannot be sure of the effects of carbon dioxide emissions, how can they determine the effects of reductions?

Most discussions of global warming focus on anthropogenic carbon dioxide, which is primarily the result of fossil fuel burning and deforestation. Since the industrial revolution, these factors have led to an increase of about 26% in carbon dioxide concentration in the atmosphere, according to the IPCC.17 Population and economic growth will greatly influence future levels of emissions. Most of today's policy discussion revolves around the following categories of mitigating the effects of greenhouse gases or adapting to climate change:

- Options that eliminate or reduce greenhouse gas emissions in order to slow or prevent warming.
- Options that offset emissions by removing gases from the atmosphere, blocking incoming solar radiation, or altering reflection of radiation.
- Options that help human society and ecosystems adapt to climate change.

Within each of these categories there are a number of possible actions with varying consequences. Some of them are very expensive, while others have potentially extreme adverse effects. Let us look at the actions being considered.

Reducing Greenhouse Gas Emissions

President Clinton's 1993 Climate Change Action Plan calls for returning U.S. greenhouse gas emissions to 1990 levels by the year 2000, in accordance with the Rio treaty, which requires a 20% reduction in emissions. What effect would this level of reduction have? In 1990 the IPCC calculated that an immediate reduction of over 60% of anthropogenic carbon dioxide, nitrous oxide, and CFCs would be required to stabilize atmospheric concentrations, along with a 15-20% reduction in methane. Clearly, a 20% reduction in overall emissions would not meet the goal of stabilizing greenhouse gas concentrations. There are various actions that can be taken to reduce greenhouse gas emissions. The methods most commonly considered include the following:

Improving Energy Efficiency and Conservation

This is no doubt the single most effective strategy, and it has additional benefits:

- Reduction of air pollution.
- Increased energy independence. (Remember the Persian Gulf War?)
- Additional time to study potential climate effects and allow for advancement of technology.
- Competitive advantage of lowered cost of manufactured products. (The United States has the highest per capita cost of energy, and this high cost is reflected in the costs of manufacturing.)

Developing Alternate Energy Sources

If reliance on carbon-containing fuels can be reduced by the development of energy sources that do not emit greenhouse gases, overall emissions should be reduced. The alternate energy sources mentioned below have all been discussed and developed to some degree.

Solar. Solar energy is one of the most promising alternative energy sources. Scientists have developed photovoltaic cells that are close to being cost competitive with fossil fuel power plants. There are currently isolated areas in which solar energy is in fact less expensive.

Wind. Windmill production enjoyed a surge during the energy crises of the 1970s. Development continued during the eighties, with smaller and smaller machines being built. According to Eliot Marshall, writing in Science, the wind energy business is booming in the nineties thanks in part to laws that encourage the use of wind-generated electricity. In 1993, there were more than 16,000 wind turbines in the United States, most of them in California. Of course, windmills require huge land areas located in areas of constant or near-constant winds.

Geothermal. Most experiments with geothermal energy sources have ended in failure. There are few reliable sources of hot springs, and many of them are very corrosive because of their mineral content. The occasional steam vents have also proven to be corrosive. There is little hope that this energy source will be exploited with a perceptible impact on the world's energy requirements.

Biomass. The development of biomass fuel sources will continue to be a slow process. Biomass fuels are all carbon-based and will have emissions of carbon dioxide associated with them. It is probable that the supply would not keep up with the demand, even if fast-growing forests are utilized for energy requirements. The costs of fossil fuels will have to be extreme before biomass will provide more than the occasional demonstration experiment.

Nuclear. Nuclear power can be and is a significant alternate energy source in many countries, including the United States. Currently, about 21% of the electric power in the United States is generated by nuclear energy; however, no new plants have been ordered since the late 1970s. The fear of nuclear accidents and the issue of radioactive wastes have made the introduction of new reactors too risky for the power utilities. In addition, the costs of building a nuclear power plant have skyrocketed largely due to government regulations and poor construction management by the utilities. When the costs of fossil fuels reach a high enough level, the world will once again turn to nuclear power in spite of its problems.

Removing Harmful Gases (Scrubbing) before They Reach the Atmosphere

The problem here, of course, is what to do with the carbon dioxide. One solution being studied is to inject it into abandoned oil or gas wells, but the cost is exorbitant and would probably raise the cost of electricity 30% or more, according to Theodore Simpson, of the United States Department of Energy.

Another solution calls for injecting CO_2 into the ocean. The earth's oceans hold many times the amount of carbon dioxide present in the atmosphere and are capable of holding much more. Unfortunately, the process of removing it from exhaust gases, compressing it, and piping it into the ocean appears to be quite expensive. In addition, there could be unwanted side effects. The carbon dioxide-rich water would be acidic and possibly deadly to organisms, and the dissolved carbon dioxide could cause seawater to lose much of its dissolved oxygen, another life-threatening effect. Ironically, federal funding for research and development in the areas of renewable energy, conservation, and nuclear fission has decreased. According to Rosina Bierbaum and Robert M. Friedman of the Office of Technology Assessment, the combined budget for these three areas was 80% lower in 1990 than it was in 1980, and bringing it up to 1980 levels would cost $2.6 billion. This expenditure would hasten the development of technologies that would reduce greenhouse gas emissions. This seems like a small price to pay for an investment that would have many benefits.

Offsetting Greenhouse Gases

Offsetting greenhouse gases after they have entered the atmosphere may be accomplished by increasing the natural carbon sink capacity of forests and by technological fixes.

Forests

Methods of modifying the carbon cycle and thereby offsetting increased carbon dioxide in the atmosphere are somewhat poorly understood. The capacity of forests to offset emissions is limited; once a tree dies or is cut down, it releases its stored carbon back into the atmosphere when it decays or is used as fuel. Even so, the deleterious effects of deforestation are generally recognized and options have been suggested for dealing with the forestry issue:

- Slow or stop the loss of existing forests. Unfortunately, large areas of land that are subject to deforestation are in tropical countries where competition for available land is intense and governmental restrictions are inadequate.
- Accelerate reforestation. George Woodwell, director of the Woods Hole Research Center in Massachusetts, calculated that two million square kilometers, or about 770,000 square miles, of new forest would be needed to remove one billion tons of carbon dioxide annually. (About three billion tons would need to be removed each year to stabilize atmospheric carbon dioxide concentrations.) Two million square kilometers represents an area the size of Texas, California, Arizona, New Mexico, and Nevada combined, or a fifth of the United States!

- Adapt forest management practices to increase the carbon stored in nonliving reservoirs, such as agricultural soils, and in artificial reservoirs, including timber products.

Geoengineering

Projects that will offset climate change through technology have been suggested for a long time. Recognition that the climate system is still poorly understood demands caution in applying these options, considering their potentially adverse effects on climate and other side effects, as well as the economic costs.

Screening Incoming Radiation. One such proposal is to shoot smart mirrors into space with rifles in order to reflect sunlight, screening it from the earth's atmosphere. Another involves putting dust or soot in orbit for the same purpose, as in a project proposed by Wallace Broecker, a geochemist at Columbia University. He suggests deploying a fleet of aircraft to load the stratosphere with sulfur dioxide. In order to offset the predicted doubling of carbon dioxide concentrations, thirty-five million tons of sulfur dioxide would need to be added per year. This could be accomplished by a fleet of 700 jumbo jets working around the clock every day, at an annual cost of about $20 billion. Those scientists who believe that we should be worrying about global cooling caused by atmospheric particulates would be quick to point out the folly of such a project.

Stimulating Cloud Formation. This project would change cloud abundance by increasing cloud condensation nuclei through carefully controlled emissions of particulate matter. The resulting clouds would both reflect and absorb solar radiation, cooling the earth. However, stratospheric particles are implicated in the depletion of the ozone layer.

Seeding the Ocean with Iron. Dubbed the Geritol solution by some, this plan involves dosing millions of square miles of ocean with iron dust, which would fertilize phytoplankton (microscopic plants). The phytoplankton, in turn, would absorb carbon dioxide, then die and sink to the bottom of the sea, storing away the carbon. While this idea worked well in theory and in small-scale laboratory tests, the practical application has been disappointing. In experiments carried out in the Pacific by marine biologists from Duke

University, the iron supplement quickly sank to the bottom of the ocean, leaving the surface waters as iron deficient as they already were. Another problem with this approach is the biological response to an increase in phytoplankton, according to an article in Chemical & Engineering News, which reported on research into iron supplementation. It turns out that an increase in phytoplankton ultimately results in an increase in the numbers of grazing zooplankton, microscopic marine animals. Because zooplankton breathe out carbon dioxide, they counteract any positive benefits of the absorption of CO_2 by phytoplankton. Geoengineering projects must be pursued with caution. Deliberately tampering with nature can have far greater consequences than the accidental effects so far created by society.

Adapting to Climate Change

The simplest, and cheapest, approach to climate change is to learn to adapt to it. Natural climate variability has always existed and will continue to do so, regardless of whether or not there are anthropogenic causes. Moreover, large numbers of people live in virtually all the earth's climate zones and move about between them. Assuming the worst effects of warming, some of the areas most affected would be agriculture, water systems, and preservation of biodiversity. The NAS study of 1991 included the following recommendations for adaptation:

Agricultural research. Agriculture is the activity most susceptible to climate change. As climate variability has always been a fact of life, agricultural practices have had to change. Special areas that researchers should explore are:

- Sustaining natural resources
- Remaining productive during extreme weather conditions
- Becoming more water efficient
- Exploiting the fertilization benefits of increased CO_2 (a point usually overlooked by global warming apostles)

Water management. Weather and precipitation cause natural variability in the water supply. Coping with this present variability can help to prepare for future climate change by increasing efficiency of water use through rights, markets, and prices, and by better management of present systems of supply.

Biodiversity. One of the factors most commonly emphasized as a consequence of global warming is an increase in the rate of loss of biodiversity. Actions recommended include:

- Establishing and protecting habitats
- Conducting an inventory of little-known species
- Collecting key organisms
- Controlling and managing wild species

As with most of the global warming issues, adapting to climate change has its advocates and its detractors. There are those who believe that the concept of adaptation will get in the way of needed action, that if we believe that adaptation to climate change is possible, the will to act politically will be diminished. Al Gore, in his book Earth in the Balance, warns that faith in our ability to adapt is a sort of "laziness in our spirit," that we have been "seduced by industrial civilization's promise to make our lives comfortable." And yet industrial civilization has done far more than make our lives comfortable. It has contributed to a healthier populace with longer life spans and more efficient use of natural resources.

Many scientists believe climate change will be so gradual that there will be ample time to adjust. One factor that is usually overlooked in the doom-and-gloom approach to making policy is the rapid advance of technology. Current predictions of temperature increase indicate that any changes will take place gradually, allowing time for adapting and for developing countermeasures, if needed. William Nierenberg, Director Emeritus of the Scripps Institution of Oceanography, points out the folly that would have resulted if the world had reacted seriously to the predictions of a 25-foot sea level rise, making unneeded engineering investments at great cost and little value. These predictions of sea level rise have steadily decreased with additional research; the current predictions are measured in inches rather than feet. In another fifteen years, will we be warned of decreasing sea levels?

Costing

Climate change has potentially enormous economic impacts. There are costs if we do nothing and warming occurs; there are costs of taking drastic action.

Costs of the Effects of CO_2 Doubling

What will be the economic effect of global warming, if it occurs? Let's look at some varying opinions of the effect on the United States.

Global warming adherents usually point to agriculture as the primary victim of climate change and warn of food shortages. Water supplies would also be affected, both for irrigation and hydroelectric power generation. Since most of the effects are predicted in terms of social costs and limited lifestyles, they are not easily quantified in dollar amounts.

However, some economic analysts have tried to put these predictions into numbers. William D. Nordhaus, an economist from Yale University, is a well-known authority on the economics of global warming. At a global warming conference in 1990, Nordhaus discussed the impact of climate change on human society and the relative economic effects of various policy responses. According to Nordhaus, the advanced countries, such as the United States, Canada, Japan, and Western Europe, have developed to the point where climate variables like temperature and humidity have little effect on economic activities, and climate is no longer much of a consideration in locating businesses. Of far greater importance are human factors, such as wages, unionization, labor-force skills, and political climate. So while the temperature variable, which is the primary focus of computer climate models, is useful as an index of climate change, the important climate variables are precipitation or water levels, extremes of droughts or freezes, and the size of projected changes as compared to day-to-day changes.

In Nordhaus 's discussion of the economic effects on the United States of a doubling of carbon dioxide (the basis used for most of the computer climate models) and a global temperature increase of 3°C, he concludes that the effects would be small around one quarter of 1% of national income. In a subsequent article, Nordhaus estimates that this would be about $15 billion at 1992 prices. Considering that the population of the United States was 249 million in 1990, the per capita cost of the effects of such global warming would be about $60 per year. Keep in mind that the IPCC 's estimate of the expected warming lies in the range of 1.5-4.5°C.

Nordhaus goes on to state that, while there is not as much statistical information available for other countries, he estimates that there would be little effect in other advanced countries. However, small countries that are heavily dependent on coastal activities, or that suffer major climate change, would be most affected. Since most poor countries rely on agriculture, the positive effects of carbon dioxide enrichment might offset damages. Those living on "the ragged edge of subsistence with few resources to divert to dealing with climate change" would be most adversely affected.

Nordhaus 's conclusion is that climate change will produce a combination of gains and losses that will not be noticeable compared to other changes, and says "those who paint a bleak picture of desert Earth devoid of fruitful economic activity may be exaggerating the injuries and neglecting the benefits of climate change." Considering the devastating impact of such uncontrollable events as earthquakes and volcanoes, the economic effects of global warming start to look somewhat negligible.

Costs of Emissions Reductions

The EPA calculated that stabilizing U.S. carbon dioxide emissions would force 30% taxes on oil and coal, whereas meeting the demands for a 20% reduction in emissions, the goal of the Climate Action Plan as well as the Rio treaty, would require a tax of $25 per barrel on oil and $200 a ton on coal. This would double U.S. energy costs.

Remember that the Department of Energy (DOE) figured the cost of a 20% reduction in carbon dioxide emissions to be as much as $95 billion per year. And according to the Office of Technology Assessment, under a "tough" scenario of reductions, which would lower emissions as much as 35% (over 1987 levels) by 2015, the United Sates could save as much as $20 billion a year or lose as much as $150 billion, based on the costs versus the fuel savings. This wide range of estimates reflects the many variables involved in making economic projections and serves as a reminder that such projections share the uncertainties exhibited by computer climate predictions. Consider the aforementioned estimates of the economic effects of global warming. A price tag of $15 billion for business as usual, as estimated by economist Nordhaus, seems small compared to a potential cost of $150 billion for lowering emissions.

Clearly, Third World countries, who are increasing their emissions at a far greater rate than the developed countries, are unable to meet such demands. The 1992 Rio treaty did not include requirements for developing countries to curtail emissions. The issue of developing versus developed countries includes more than just relative effects of emissions on the atmosphere, as we shall discuss shortly.

Carbon Tax

A common solution that is frequently bandied about is the carbon tax. Georgii S. Golitsyn, of the Moscow Physical Technical Institute, put it this way at the Greenhouse Glasnost, a conference organized by Robert Redford through his Institute for Resource Management: "I propose taxing every ton of fossil fuel consumed in developed nations at a small percentage of its cost. From these proceeds an international assistance fund could be established for developing nations to introduce energy- and resource-saving technologies." Sounds good, but carbon taxation is a very complicated issue. How it is to be applied internationally, whether or not the taxes are applied as production or consumption taxes, who should get the revenues, and how the carbon content is calculated are all questions that contribute to a problem requiring a great deal of study. Nevertheless, taxing coal-based fuel usage could have the effect of encouraging a shift to renewable fuels as well as energy efficiency and conservation.

Political Sensitivity

The most common response to the position that burning of fossil fuels contributes to global warming is to call for cutbacks in their use (and in fact that approach is included in the 1993 Climate Change Action Plan, which calls for returning U.S. greenhouse gas emissions to 1990 levels by the year 2000.) But the buildup of carbon dioxide is a global phenomenon because of atmospheric circulation, and if emissions are to be controlled, the effort will have to be made on a worldwide basis.

And although it is a global problem, it is not an equitable one. All nations are not equally responsible. In fact, as industrialized countries have economized on labor and capital by shifting away from coal and toward oil, gas, and carbon-free fuels, their per capita carbon dioxide emissions have been reduced, while developing countries have increased their carbon dioxide emissions. Note that while the carbon dioxide emissions of the developed nations have leveled off or even decreased, the developing nations of India and China are increasing both their total and their per capita emissions. The developing nations are also increasing their populations at much faster rates than the developed nations. These trends should be expected for all the developing countries.

All nations are not able to cut emissions equally, and the developing countries would need a great deal of assistance to do so and still continue to develop. All nations are not equally affected by climate change. Some would actually benefit from increased rainfall and a more temperate climate, while others would suffer from coastal erosion and flooding.

Fred Singer, well-known climatologist, pointed out the realities of these inequities in 1989, when he wrote that because fossil fuel is a major economic commodity, with tremendous investments already incurred by the majority of countries, it would be economically disastrous to abandon it as an energy resource. Since carbon dioxide buildup is a global phenomenon, any course of action would have to be agreed upon and implemented internationally But because the perceived climate change will actually be advantageous for some areas, those areas might be reluctant to make sacrifices. Since the scientific community is still in disagreement about the future climate, many people are reluctant, and rightly so, to take drastic action that would have a profound effect on the economy.

There is also the sensitive issue of resentment, on the part of developing countries, of the developed countries who have had the advantages of an increased standard of living because of the very activities that created the problem, and who wish to deprive developing countries of the same benefits. As pointed out in a 1989Forbes article, "there is no way to prevent a [Third World] CO_2 doubling without slashing growth and risking a revolt of the have-not nations against the haves." This reality was reinforced at Rio. In a 1992 article in the "San Francisco Chronicle", Maurice Strong described criticism hurled at the industrialized countries by a negotiator from Malaysia, Wen Lian Ting.Wen insisted that unless rich countries pay Malaysia not to cut its trees the cutting will continue. Malaysia is not going to keep its trees "in custody for those who have destroyed their own forests and now try to claim ours as part of the heritage of mankind!"

An editorial in Nature reminds us: "Rich countries in the nineteenth century were called imperialists because of their fondness for telling the poor how to run their affairs."[39] Certainly the industrialized countries are in a tenuous position when it comes to making environmental demands on developing countries.

Changing U.S. Political Climate

The political climate has, of course, changed with each presidential administration. We've already mentioned Vice President Gore's response to scientific uncertainties. While a United States Senator, Gore was on the Committee on Commerce, Science and Transportation. In the 1991 hearing on the NAS's report "Policy Implications of Greenhouse Warming," Gore criticized the report (and the scientists) for being overly cautious. Frank Press, president of NAS, responded:

Sometimes it is not a question, it seems to me, of being cautious; it is a question of what makes good sense based on scientific evidence. Sometimes we get ahead of our science. I suspect the Soviets thought when they built Chernobyl that they were in full grasp of all of the science necessary to build and effectively run that plant. They were not. So, you can sometimes go too fast, and too far, just as you can sometimes be too cautious and too slow.

It is not an easy task to pick the right path.

Press's comments serve as a chilling reminder of the events of Chernobyl and their effect on the fears of the use of nuclear energy. The lesson, of course, is not that nuclear energy is to be feared, but that proceeding too fast in the face of uncertainties is risky.

The IPCC also finds itself in a politically sensitive situation because of its influence on policymaking. Some scientific journals and the popular press have criticized IPCC conclusions "for lack of openness about uncertainties and for brushing aside controversies." While we have found the reports to mention the uncertainties, we have found the statements couched in qualifiers. And it is somewhat difficult to reconcile the fact that IPCC states the uncertainties while still drawing the conclusions that appear to be politically correct, conclusions that recommend taking drastic action.

Flying Too Close to The Sun

Remember Daedalus and Icarus.The hysteria of the global warming myth threatens to be the fuel that will fly us too high, too soon. In 1984, Bernard Cohen, a physicist at the University of Pittsburgh, warned of the dangers of scientific policy being influenced by the emotional responses of the public when he said:

Our government's science and technology policy is now guided by uninformed and emotion-driven public opinion rather than by sound scientific advice. Unless solutions can be found to this problem, the U.S. will enter the 21st century declining in wealth, power and influence. The coming debacle is not due to the problems the environmentalists describe, but to the policies they advocate.

Cohen was rightly concerned. As we have seen, public policy is being increasingly influenced by the loud cries of environmental activists and the scientists who strongly support the global warming theory.

The warnings continue. A 1994 editorial in Science warns of the power of environmental organizations:

Concern continues that the U.S. Environmental Protection Agency (EPA) might curtail or ban the production of chlorine and compounds containing it. This perception has been fostered by indications that EPA policy is being predominantly influenced by Greenpeace and its allies.

Remember that the EPA was established as an agent of the federal government to set air quality standards and demand compliance, and it is possibly the most powerful governmental agency on earth. Careful consideration must be given to any policy that will give it more power, as would happen with more stringent emission standards.

The foundation on which the global warming theory is built is full of inconsistencies and inadequacies. It is imperative that the uncertainties in the science of the global warming theory be cleared up before irreversible policy decisions are made.

11

Possible Solutions of Global Warming

Introduction

Addressing the consequences of global warming will demand, on a worldwide scale, the kind of social and economic mobilization experienced in the United States only during its birthing revolution and World War II, and therein lies a problem. The buildup of greenhouse gases in the atmosphere is a nearly invisible, incremental crisis. Carbon dioxide is not going to bomb Pearl Harbor to kick start the mobilization. Author Jonathan Weiner observes, "We do not respond to emergencies that unfold in slow motion. We do not respond adequately to the invisible".

It has been said (not for attribution) that the best thing which could happen to raise worldwide concern about global warming would be a quick collapse of the West Antarctic ice sheet, which would raise worldwide sea level a notable number of feet over a very short time. When stock brokers' feet get wet on the ground floor of New York City's World Trade Center, all the world's competing economic interests might mobilize together and provide the sociopolitical responses necessary to address the atmosphere's overload of greenhouse gases before it is too late. The same water that could lap at the ground floor of the Trade Center also would ruin most farmers in Egypt and Bangladesh and slosh in the lobbies of glass towers of Hong Kong and Tokyo. Perhaps, only then, might all of humankind heed the implications of Chief Seah'tl's farewell speech a century and a half ago. We may be brothers (and sisters) after all. We shall see.

This is not a secret crisis, just a politically unpalatable one. Al Gore, in Earth in the Balance: Ecology and the Human Spirit, raised a sociopolitical call for mobilization against human-induced warming of the Earth: "This point is crucial. A choice to 'do nothing' in response to the mounting evidence is actually a choice to continue and even accelerate the reckless environmental destruction that is creating the catastrophe at hand". In his book, Gore, then a U.S. senator, called for a "global Marshall Plan," to include stabilization of world population, the rapid creation and development of environmentally appropriate technologies, and "a comprehensive and ubiquitous change in the economic 'rules of the road' by which we measure the impact of our decisions on the environment".

Eight years after Gore issued his manifesto, fossil-fuel emissions had risen in the United States. Gore had captured the Democratic Party's nomination for president of the United States, and global warming had slipped from campaign radar. From this vantage point, one imagines the world lurching through the twenty-first century as global public opinion slowly galvanizes around year after year of high temperature records, and as public policy only slowly begins to catch up with the temperature curve. The temperature (and especially the dew point) may wake the global frog before he becomes poached meat. Whatever the outcome of the public policy debate, the odds are extremely high that the weather of the year 2100 will be notably warmer than today, as greenhouse "forcing" exerts an ever-stronger role in the grand dance of the atmosphere which produces climate.

Ross Gelbspan observes,

Global warming need not require a reduction of living standards, but it does demand a rapid shift in patterns of fuel consumption reduced use of oil, coal, and the lighter-carboned natural gas to an economy more reliant on solar energy, fuel cells, hydrogen gas, wind, biomass, and other renewable energy sources. It is doubtful that capitalistic market forces will bring about this shift on their own, because market prices of fossil fuels do not incorporate their environmental costs.

George Woodwell has been quoted as saying, "For all practical purposes, the era of fossil fuels has passed, and it's time to move on to the new era of renewable sources of energy." The other alternative, says Woodwell, is to accept the fact that "the Earth is not simply moving toward a new equilibrium in temperature. It is entering a period of continuous, progressive, open-ended warming".

In Jeremy Leggett's opinion,

The uniquely frustrating thing about global warming to the many people who see its dangers is that the solutions are obvious. There is no denying, however, that creating the necessary changes will require paradigm shifts in human behavior particularly in the field of cooperation between nation-states which have literally no precedent in human history. There is no single issue in human affairs that is of greater importance.

According to a Greenpeace Report edited by Leggett, "The main routes to surviving the greenhouse threat are energy efficiency, renewable forms of energy production less greenhouse-gas-intensive agriculture, stopping deforestation, and reforestation". Greenpeace also recommends redirecting spending away from armaments and toward development of a sustainable energy for the future of humankind.

Of the broader picture, Michael MacCracken writes,

The underlying challenge is for industrialized society to achieve a balanced and sustainable coexistence with the environment, one that permits use of the environment as a resource, but in a way that preserves its vitality and richness for future generations. The challenge is to transform our ways before the world is irrevocably changed toward displacing militarization and the ever-increasing push for greater national consumption as the primary driving forces behind industrial activity.

According to Donald Goldberg and Stephen Porter of the Center for International Global Law:

The Clinton administration has bungled repeated chances to initiate domestic measures. For example, recent legislation proposed by the White House to restructure the electric utility industry could have been crafted to require utilities to reduce their carbon-dioxide emissions. In fact, the Environmental Protection Agency lobbied hard for the authority to impose a cap-and-trade program on utilities' CO2 emissions, similar to the trading system that has lowered sulfur dioxide (SO2) emissions in a cost-effective way. This was a golden opportunity, as the restructuring bill is projected to save the average consumer roughly $200 a year, which would have more than offset the cost of reducing GHG [greenhouse-gas] emissions. Unfortunately, the White House chose to forgo this opportunity.

According to Goldberg and Porter, loopholes in the Kyoto Protocol, adopted at the insistence of the United States, permit richer countries to avoid many of its mandated emission reductions by purchasing allowances from other countries through the protocol's "flexibility mechanisms." The Buenos Aires Climate Conference negotiated a mechanism allowing trade in greenhouse gas emission rights in two markets. The first market would allow "sellers," nations which exceed greenhouse gas-reduction targets set in the 1997 Kyoto Protocol, to offer their rights for sale to "buyers," countries which have not met their targets. The second market, the Clean Development Mechanism, will allow industrialized countries to meet part of their greenhouse-gas-reduction quotas by transferring clean technology to poorer countries so that antipollution projects can be carried out there.

"If it buys all (or most) of its reductions," Goldberg and Porter write, "the United States will not get its own house in order. In the long run, efficiency and productivity in the U.S. economy will suffer because domestic industry will be shielded from any incentive to adapt". Under these provisions, the United States could "purchase" emission reduction credits from nations, such as Russia and Ukraine, which reduced their greenhouse-gas emissions during the 1990s because their economic infrastructure collapsed.

The continuing political wrangling over the Kyoto Protocol illustrates why the world is responding so slowly to the impending crisis of global warming. Climate diplomacy remains an arena dominated by competition of special (mainly national) interests. Meanwhile, a few countries, most of them in Europe, are taking steps to mitigate greenhouse forcing on their own. While British emissions of greenhouse gases by the year 2000 had fallen between five and six percent compared to the Kyoto Protocol 1990 targets, emissions in the United States rose 11 percent between 1990 and 1998. Canada's greenhouse-gas emissions rose 13 percent during the 1990s, while several European countries (including Britain) made substantial progress toward meeting the goals of the Kyoto Protocol by reducing their greenhouse-gas emissions as much as 10 percent compared to 1990 levels.

Denmark (which produces less than one percent of humankind's greenhouse gases) underwent something of a mobilization against global warming during the 1990s. Denmark was planning "farms" of skyscraper-sized windmills in the North and Baltic seas that, if plans materialize, will supply half the nation's electric power within 30 years. The Danish wind-energy manufacturers' association believes that electricity produced through wind power on a large scale will be financially competitive with power from plants burning fossil fuels, which will be phased out if wind power proves itself. Svend Auken, Denmark's environmental and energy minister, said that with half of his country's power coming from Norwegian hydroelectric plants and the other half from wind power, the country is planning to meet its electricity needs within three decades while reducing carbon dioxide production to nearly zero. The wind farms must prove their endurance in winter storms and stand up to the corrosion of seawater, but if they can, Denmark's windmills will prevent the production of 14 million tons of carbon dioxide a year.

While the fossil-fuel economy remained firmly entrenched in most of the world at the turn of the millennium, gains were being achieved in some basic areas of energy conservation. In 1994, for example, the average person in the United States was recycling 380 pounds a year, up from 62 pounds in 1960, a 613 percent increase. Following the passage of the Clean Air Act in 1972, the United States also made a concerted effort to limit the production of nitrous oxides by gas turbine engines. Before regulation, the typical gas turbine engine emitted 200 parts per million (p.p.m.). Since then, several technological innovations have reduced emissions to below 10 p.p.m. Technology was being developed in the late 1990s which could reduce the rate to two to three p.p.m.

IPCC's Coping Strategies

The Intergovernmental Panel on Climate Change's (IPCC) Second Assessment describes technological coping strategies which may help reduce the rapid rise in atmospheric greenhouse gases. For example, according to the IPCC, gains in energy efficiency of 10 to 30 percent "are feasible at negative to zero cost in each of the sectors in many parts of the world over the next two to three decades". With more attention to technological improvements, according to the IPCC, efficiency gains could reach 50 to 60 percent in some industries. However, according to the IPCC, "Because energy use is growing worldwide, even replacing current technology with more efficient technology could still lead to an absolute increase in greenhouse-gas emissions in the future". The IPCC recommends a wide-ranging, intense examination of transportation, industrial processes, and other human activities which produce greenhouse gases with energy conservation and efficiency in mind. The recommendations extend to alternative, "clean" methods of energy production (biomass, solar, wind, etc.) and to management of the world's forest cover with its ability to absorb carbon dioxide in mind.

A number of measures could conserve and sequester substantial amounts of carbon over the next 50 years. In the forestry sector, measures include sustaining existing forest cover; slowing deforestation; natural forest regeneration; establishment of tree plantations; promoting agroforestry.

The IPCC's Second Assessment also examines the use of taxes on carbon emissions, as well as on the production and use of energy derived from fossil fuels, encouragement of research and development into alternative energy technologies, along with "phasing out of existing distortionary policies which increase greenhouse-gas emissions, such as some subsidies and regulations".

The world economy and indeed some individual national economies suffer from a number of price distortions which increase greenhouse-gas emissions, such as some agricultural and fuel subsidies and distortions in transport pricing. A number of studies of this issue indicate that global emissions reductions together with increases in real incomes are possible from phasing out fuel subsidies.

The IPCC's estimates of climate change in the next century are based, for the most part, on conservative assumptions that do not include such factors as possible sea level rise from partial disintegration of the West Antarctic Ice Sheet, or major changes in ocean circulation. Even using conservative projections, the tone of many IPCC reports is sobering. The IPCC's outlook for countries such as China and India, with large populations and rapidly growing resource-consuming middle classes, indicates that achievements in energy efficiency and development of fuels which do not consume fossil fuels "are likely to be insufficient to offset rapidly increasing emissions baselines, associated with increased economic growth and overall welfare".

Carbon as a Taxable Commodity

As early as the 1980s, Colorado Senator Timothy Wirth introduced legislation in the United States Senate Committee on Energy and Natural Resources to place mandatory controls on industries which produce greenhouse gases. The proposal was received by lobbyists (the centurions of the special-interest state) like a proverbial lead balloon, as Senator Wirth's measures languished in committee. Wirth also sought to legislate stringent regulation of energy efficiency, especially in automobiles, which produce 40 percent of human-generated carbon dioxide in the United States. Wirth favored limits on population growth and increased governmental support for alternative energies (wind, solar, hydrogen) which do not burn fossil fuels.

Several economists have suggested that greenhouse-gas emissions be taxed. As of 1990, however, carbon taxes were still a matter for economic theorists in all except a few European nations: Sweden (about $62 a ton), Finland ($6.50 per ton), and the Netherlands ($1.50 a ton).

Chris Flaven of the World watch Institute has proposed a worldwide carbon tax of $50 a ton. He also proposes paying 10 percent of the tax into a fund to subsidize development of new technologies that will reduce emissions of greenhouse gases. A study of greenhouse-gas emission-reduction strategies in Chicago supported a "CO2 Fund which would pool resources from the state and private industry and make funds available in low-income communities in order to encourage emissions reductions". In Earth in the Balance, Gore supports a fossil-fuel tax, part of which would be placed in an Environmental Security Trust Fund, "which would be used to subsidize the purchase by consumers of environmentally benign technologies, such as low-energy light bulbs or high-mileage automobiles".

William D. Nordhaus has modeled the economic effects of carbon taxes ranging from $5 to almost $450 per ton. Nordhaus provides a range of carbon-tax rates with expected reductions in fossil-fuel use and costs to gross national product in the United States: he expects a carbon tax of $13 a ton to reduce carbon emissions six percent (while reducing greenhouse gases 10 percent), at a negligible cost to Gross National Product (GNP). A carbon tax of $98 a ton is expected, by Nordhaus, to reduce greenhouse gases 40 percent, with a half a percentage point decrease in GNP. Nordhaus' highest tax estimate, $448 per ton, would decrease greenhouse gases 90 percent, and lower GNP by more than four percent.

Nordhaus estimates that a $5 per ton tax would raise the price of coal 10 percent, the price of oil 2.8 percent, and the price of gasoline 1.2 percent. Applied worldwide, the same level of carbon taxation would, according to Nordhaus' models, reduce greenhouse-gas emissions 10 percent, provide $10 billion in tax revenue, and add $4 billion per year to the global economy. A $100 per ton tax on carbon emissions would raise the price of coal 205 percent, the price of oil 55 percent, and the price of gasoline between 23 and 24 percent, according to Nordhaus. The $100 tax would reduce greenhouse-gas emissions by an estimated 43 percent (close to the level recommended by the IPCC to forestall significant global warming). A $100 tax would provide $125 billion in tax revenues and would decrease global "net benefits" $114 billion, according to Nordhaus.

Michael E. Mann and Richard J. Richels contend that a carbon tax of $250 a ton will be necessary to suppress carbon dioxide emissions by 20 percent. Such a tax would add about 75 cents to the cost of a gallon of gasoline, and $30 to the cost of a barrel of oil.

More than 2,000 economists, including six Nobel Laureates, endorsed a statement on global warming during 1997 which called for the use of market mechanisms, including carbon taxes, to move world economies away from fossil fuels:

As economists we believe that global climate change carries with it significant environmental, economic, social, and geopolitical risks, and that preventive steps are justified. Economic studies have found that there are many potential policies to reduce greenhouse-gas emissions for which the total benefits outweigh the total costs. For the United States in particular, sound economic analysis shows that there are policy options that would slow climate change without harming American living standards and these measures may in fact improve U.S productivity in the longer run.

The United States tax code has been suggested as a tool to reduce fossil-fuel emissions, with such proposals as tax credits for energy-efficient appliances, changes in building codes focused on energy efficiency; energy efficiency standards for universities and health centers which receive federal grants, as well as tax incentives for fuel-efficient cars.

Stephen Schneider observes that any significant carbon tax probably would shape the future course of technology:

When the price of conventional energy goes up and will stay up because everyone knows the carbon tax or quota system will be here to stay then a host of entrepreneurs and governmentally assisted labs will take up the challenge to develop and test an unimaginable array of more efficient and decarbonized production and end-use alternatives. The higher price of carbon-intensive fuels will spur the Research and Development investments, which economists call induced technological changes.

Full Cost-Pricing of Energy

A 1991 National Academy of Sciences report recommended a "study in detail of the 'full social cost pricing' of energy, with a goal of gradually introducing such a system". Such a system would price energy to include not only the costs of production and distribution, but environmental costs as well. The study admits that "such a policy would not be easy to design or implement". Such a change would involve a full-scale restructuring of the accounting rules in the capitalistic marketplace, which defines "cost" only in terms of financial (not environmental) assets.

Schneider explains full cost-pricing of energy under the rubric of "integrated assessment."

Integrated assessment is an attempt to merge economy and ecology with the inclusion of externalities and the analysis of end-to-end costs. In other words, the price of a lump of coal isn't simply extraction, storage and transport, but health consequences of mining and burning, as well as the whole range of potential environmental alterations from the production and use of energy. The cost of a car isn't simply the materials, labor and profit, but should also include disposal costs, and tailpipe emissions. Integrated assessment will fall short on several grounds because of the technical difficulty of evaluating the whole range of costs and benefits of our activities and because, ultimately, the single unit of comparison ismonetary currency.

Thomas R. Casten endorses an end to annual federal government spending of $4 billion on subsidies for producers of fossil fuels in the United States, mainly from tax credits related to depletion allowances. Such a change could take place under present accounting rules and assumptions. In the longer term, several observers, including Al Gore, propose that "the definition of Gross National Product (GNP) should be changed to include environmental costs and benefits," and that "the definition of productivity should be changed to reflect calculations of environmental improvement or decline." Gore calls such definitions "Eco-nomics".

The Sierra Club also asserts that subsidies for fossil-fuel exploration should be ended. Instead, according to the Sierra Club, emissions of pollution should be taxed, as government increases research and development expenditures for solar and other renewable-energy technologies. According to a Sierra Club study, the cost of generating wind energy fell 85 percent between 1981 and the middle 1990s. Improvements in photovoltaic cell technology also have reduced the per unit cost of generating solar power. Despite the growing competitiveness of alternative fuels, Gelbspan argues that 90 percent of U.S. energy subsidies and incentives still go to fossil fuels, compared to 10 percent for energy alternatives which may help slow the rise in atmospheric greenhouse-gas levels.

Higher Rates of Energy Efficiency

A number of reports have recommended that all energy use be closely scrutinized for technological improvements. Electricity-generating industries, automobiles, the home, and so forth have been studied for ways to save energy. Many observers believe that at least 50 percent of present energy use could be saved through changes that are available through existing technology. Only about one-third of the energy consumed by a typical coal-burning power plant in the United States actually drives steam generators which produce electricity. The other two-thirds of the energy consumed by a typical coal- or oil-fired power plant is lost as waste heat through smokestacks and cooling towers. Increasing attention is being paid to "co-generation" strategies that capture more of that waste heat for actual energy production. Industry in the United States made 25 percent more efficient use of energy in 1986 than in 1973.

Existing power plants offer many opportunities for conservation: "For a pulverized coal plant, 90 percent of the carbon dioxide can be recovered using a chemical absorption process to clean up flue gases". In such a process, according to C.A. Hendriks and colleagues of the Netherlands' University of Utrecht, "The carbon dioxide is recovered by leading the flue gas through a solution containing the absorber". The cost of removing 90 percent of the carbon dioxide is roughly a 25 percent loss in generation efficiency and roughly a 35 percent increase in the cost of electricity, according to Hendriks.

According to many scenarios, coal- and oil-fired power plants would first be streamlined to generate electricity as efficiently as possible, then phased down, and finally out, in favor of yet-to-be-developed electricity generation capacity derived from biomass, solar photovoltaics, solar thermal, and wind. Carbon dioxide generated by fossil-fuel burning plants would be buried in the deep ocean or under depleted oil and gas fields.

Energy conservation strategies produced some gains in energy efficiency in the United States following the oil-shortage scares of the 1970s. In 1973, roughly 18,000 British Thermal Units (BTUs) of energy was required to produce a dollar of gross domestic product in the United States. By 1999, roughly 11,000 BTUs was required to produce the same amount (in constant-dollar terms), a decline of roughly 30 percent. In the United States, oil expenditures amounted to 8.5 percent of gross domestic product in 1981, and only 3 percent in 1999, according to the U.S. Energy Department.

Basic Changes with Existing Technology

Necessary changes could begin incrementally, on an individual basis, with low-technology, close-to-home measures such as painting buildings light colors, replacing standard light bulbs with more energy-efficient types, and requiring significantly higher gas mileage from all forms of motorized transportation.

Eventually, however, human economic activity and energy requirements must be decoupled as much as possible from the fossil fuels that are raising the levels of atmospheric carbon dioxide and other greenhouse gases.

A National Academy of Sciences report, issued in 1991, considers greenhouse-gas mitigation strategies in some detail and concludes that basic energy-conservation strategies (many requiring technological upgrades of infrastructure) could lower greenhouse-gas emissions in the United States by roughly one-third from 1990 levels. This report contains a detailed list of such strategies, including many basic conservation measures in existing homes, businesses, cars, and trucks.

The National Academy of Sciences' list of mitigation strategies starts with the most prosaic a reduction of air-conditioning usage and the urban "heat island" effect through a crash program to paint half the roofs in urban areas white. Next, the report recommends another crash program to replace incandescent lighting with compact fluorescent bulbs. Energy-efficiency technology would then be upgraded in residential and commercial water heaters, commercial lighting, commercial cooking, refrigeration, and appliances. Homes would be heavily insulated to conserve fuel consumed by space-heating and air-conditioning. Industrial energy efficiency would be upgraded through the use of co-generation, fuel switching, and new technology.

The American Council for an Energy-efficient Economy estimates that 35 to 40 percent of the electricity consumed in the United States could be saved through existing technology and rigorous conservation measures. Electricity consumed by lighting could be cut nearly in half through adoption of compact fluorescent light bulbs. These bulbs consume one-sixth the energy used by "standard" bulbs. Singapore, the Philippines, Indonesia, Malaysia, and Thailand reduced their energy consumption by more than 20 percent in a decade (1985–1995) by implementing more efficient building codes. The Marshall Islands began installing solar hot-water heaters to replace oil-fueled machines used by many of its 55,000 residents.

Some proposed changes are more behavioral than technological such things as building more bicycle paths and urging workers to "telecommute" when possible. Other observers advocate dietary measures, such as avoiding beef in favor of plant-based foods. Cattle are responsible for 72 percent of methane released by livestock in the United States. Consideration has been given to farming with an eye to reducing greenhouse gases. Under this system, traditional farm machinery is not abandoned, but used less often. Natural sources of fertilization (such as legumes) are substituted when possible for synthetic fertilizers. Such a farm is designed as an ecosystem in which many wastes are recycled. Many of the practices of "sustainable agriculture" (such as spreading manure on fields) are hardly new. Some of them recall farming methods of pre-industrial times.

A.B. Lovins and L.H. Lovins comment,

Organic farming techniques are already rapidly spreading for economic, health, and environmental reasons [which] can simultaneously reduce biotic CO2, N2O [Nitrous Oxide], and CH4 [Methane] emissions, directly from farmland, and indeed can reverse the CO2 emissions.

A study by the Illinois Environmental Protection Agency found that a number of basic energy-saving measures (including heating-system retrofits, improved thermostats, better efficiency in hot-water generation, more insulation, installation of storm windows, and lighting retrofits) could reduce energy use in an average home by 22 percent. The same study also found that similar efficiencies in public buildings could reduce their natural-gas consumption by 20 percent, and electricity consumption by 13 percent.

Lovins and Lovins caution that the infrastructure changes required to retool machinery to use energy more efficiently will be expensive: roughly $200 billion a year for the United States, in 1989 dollars. Lovins and Lovins argue against the claim that mitigating global warming will not drastically curtail American lifestyles.

Nothing could be further from the truth. The fuel-saving technologies that can stabilize global climate while saving money actually provide unchanged services: showers as hot and tingly as now, beer as cold, rooms as brightly lit homes as cozy in the winter and cool in the summer. The quality of these and other services can often be not just sustained but substantially improved by substituting superior engineering for brute force.

In Japan, by the beginning of 1999, control of greenhouse-gas emissions had reached the local (or prefecture) level. On April 1, Japan enacted a stringent law to combat global warming. The law is meant to bring Japan into compliance with the 1997 Kyoto Protocol on climate change. Each municipality has been instructed to submit its own plan to bring down greenhouse-gas emissions.

The city of Koga, in Ibaraki Prefecture, during July 1998, began to actively promote the bicycle as a means of daily transport. Two months earlier, the city hosted a conference among local governments from throughout Japan. Participants traveled between conference sites by bicycle. The city published a pamphlet offering hints on how to best use bicycles, and purchased twenty bikes for official use. Cyclists have been given the right-of-way on some roads in the city.

The San Francisco Board of Supervisors passed a resolution in 1998 supporting the Kyoto Protocol on climate change. The city of Oakland quickly followed.

Whereas, the City of San Francisco has begun to address its local contribution to global climate change through:

1. Setting a long-term goal of eliminating climate-changing and ozone-depleting emissions and toxics associated with energy production and use, as set forth in the Sustainability Plan adopted in 1997,
2. Passage of Resolution #227–95, supporting San Francisco's Participation in the Cities for Climate Protection Campaign, in which it pledged to take a leading role in reducing energy consumption, establishing a greenhouse gas reduction goal, and developing a local action plan to reduce local greenhouse gas emissions. Therefore, be it resolved that the City and County of San Francisco supports the Kyoto Protocol on Climate Change as a small but significant means to reduce greenhouse gas emissions and stabilize the global atmosphere

and as a necessary first step toward maintaining the health and quality of life for future generations.

During 2000, Oakland, California, approved a plan to buy alternative power for all of its municipal needs. "It leads us in the direction of reducing global warming, stimulating new industry, and sets the pace for the national government," said Oakland Mayor Jerry Brown, a former California governor and presidential candidate. Santa Monica last year became the first city in California to purchase green power for all its municipal needs during 1999. Palmdale soon followed and several other communities were preparing to do the same during 2000, including Santa Barbara, San Jose, and Santa Cruz. By the year 2000, one-eighth of the power consumed in California came from non fossil-fueled sources, which include wind, geothermal, solar, methane, and hydropower from small dams. Large hydroelectric plants are not considered environmentally friendly.

Urban Zoning Changes

Even as climate scientists rang alarm bells regarding global warming during the 1990s, the suburban rings of most major urban areas in the United States continued to grow, intensifying the reliance of many residents on their automobiles, which often were being driven longer distances to and from work and other engagements, raising greenhouse-gas emissions. A number of observers have recommended denser urban development that is more amenable to mass transit and walking. A group of scientists led by S.T. Boyle graphed gasoline consumption against urban density. Not surprisingly, the greatest energy consumption occurred in the least dense, newer urban areas in the United States (Houston, Denver, Los Angeles, etc.) which expanded rapidly after the automobile diffused early in the twentieth century. The densest (low-consumption) urban areas were in Europe and Asia, with large areas of urban infrastructure predating the automobile, such as London, Hamburg, and Tokyo.

"A Law of The Air"

The phrase "law of the air" was first used by Margaret Mead during a meeting in the early 1970s and repeated in "The Atmosphere: Endangered and Endangering," a Fogarty Conference report that she and William Kellogg edited. Stephen Schneider picked it up in his works during the 1980s. "Of course, we now have a 'law of the air' known as the Kyoto Protocol," commented Schneider. Thus, we now have a design for a "Law of the Air." We will have a real Law of the Air when countries and companies can be taken to court, convicted, and penalized for illegal emission of greenhouse gases.

Change the Automobile

In a laboratory near Stuttgart, Germany, engineers from Daimler Chrysler have been developing an experimental motor vehicle which will run on a hydrogen fuel cell, emitting only water vapor from its exhaust pipe. Fuel cells of this type have been used by NASA to generate power aboard space vehicles. Only during the 1990s, however, did hydrogen fuel cells become small enough to use as automobile engines. The prototype NECAR4 is modeled on a Mercedes sedan and is capable of carrying five people and luggage 280 miles between fueling stops at speeds of up to 90 miles (145 kilometers) per hour. Similar vehicles may be on the market by the year 2005. In any fuel cell, electricity is required to separate water into hydrogen and oxygen atoms. Some-day, this electricity may be produced by solar and wind power, but with existing technology, such generation would be too expensive to make hydrogen a competitive fuel. The first hydrogen-fueled cars probably will be charged with electricity from existing sources, including fossil fuels.

Less than 20 percent of the energy consumed by a typical Ford Escort (an "economy car") is used to move the NECAR4. The rest is discharged, largely as waste heat, through the car's radiator and tailpipe. Late in the twentieth century, one-fourth of the world's greenhouse gas total was being produced by more than 600 million automobiles which are, for the most part, generally as inefficient as that Ford Escort.

Proposals have been aired to steadily change the technology and fuel sources of automobiles to reduce greenhouse emissions. Under most of these plans, automakers would be required to produce more efficient internal combustion engines. As technology becomes available, automobiles would be adapted to operate on "hydrogen, electricity, and carbon fuels from biomass sources". By the year 2100, the gasoline-burning automobile of the twentieth century may be a museum piece. Combustion of fossil fuels may seem as old-fashioned to people of that time as a horse and buggy appears today.

According to a study by the National Academy of Sciences (NAS), released in 1991, greenhouse-gas emissions could be reduced markedly through increasingly higher mileage requirements for private automobiles and heavy trucks. The NAS study proposes that mileage standards should rise to about 48 miles per gallon (m.p.g.) for private vehicles and 40 m.p.g. for heavy trucks. People who commute in an automobile alone also would be severely taxed to encourage what the report calls "transportation-demand management". The Sierra Club recommends raising Corporate Average Fuel Economy (CAFE) standards in the United States to 45 m.p.g. for cars and 34 m.p.g. for light trucks by 2005 as "the biggest single step the U.S. can take to curb global warming and reduce our dependence on oil". Patricia Glick comments, "By simply adding existing technology to their vehicles, automobile manufacturers can slash global warming pollution and save consumers money at the same time". The Sierra Club points out that an automobile which meets its mileage standards was being manufactured during the 1990s.

During 1997, Toyota introduced the world's first hybrid gas-electric car, the Prius, with much lower greenhouse-gas emissions than conventional automobiles. Marketed as a "green" sedan, the Prius has since sold so quickly in Japan that Toyota has opened a second production plant. By the year 2000, Honda was selling the Insight, a hybrid gasoline-electric car that gets 61 miles to a gallon of gasoline in start-and-stop city driving and 70 m.p.g. on the open road. Such cars were capturing only a small fraction of the market in the United States, however. Honda's initial shipment of Insights to the United States totaled only 4,000 cars.

In Detroit, General Motors President John Smith said in 1998, "No car company will be able to survive in the twenty-first century by relying on the internal-combustion engine alone". Early in 2000, General Motors unveiled a five-passenger, 54-horsepower automobile that it said will have a fuel efficiency of 80 miles to a gallon on the open road. The Precept uses a small diesel engine with an electric motor. One problem cited by engineers on these cars is that the diesel engine produces more nitrous oxides than air-pollution standards allow. The Precept's engine is in the rear of the car (to improve airflow). The car cuts weight by using aluminum and titanium rather than steel for some parts. This car is similar in size to a Chevrolet Malibu, but weighs 460 pounds less.

One problem with technological fixes is that humankind's mechanical infrastructure does not refashion itself instantly, and the poorer a region, the longer the transition. Automobiles are a very good example of this delay. Fuel standards mandated today will not apply to the entire fleet until older vehicles are retired from the road. A report on reducing greenhouse-gas emissions in Chicago endorsed "clunker" trade-in programs, which help to reduce greenhouse-gas emissions by paying nominal amounts of cash to retire "high emitters" from the road. Several clunker-purchase programs have been implemented in California, Delaware, and Illinois. The Illinois Environmental Protection Agency implemented a clunker trade-in program that targeted vehicles built before 1980. The cost per vehicle ranged from $647 to $950 depending on its age. The average cost was $860 per vehicle.

Improved fuel efficiencies may reduce greenhouse-gas emissions for the average individual automobile, but these savings are being more than outweighed by increasing numbers of cars on the road. The world fleet of automobiles and light trucks was 53 million in 1950 and 400 million by 1990. Annual production was 10 million in 1950 and 50 million in 1990. A report by the World Resources Institute, published in 1990 asserts that fundamental changes in transportation technology will be required to stabilize or reduce greenhouse gases. Hydrogen-powered and electric cars are mentioned, but the report does not speculate on how the electricity that would power these vehicles should be generated.

Wind Energy and Other Alternative Sources

Some major energy and natural resource companies were developing large scale wind-power projects by the late 1990s. One example was Northern Alternative Energy's $32 million project, which entails construction of two wind-energy projects with Northern States Power (NSP) Company. The two projects will add a combined installed capacity of 23 megawatts to NSP's wind-energy resources. The new projects will be located on Buffalo Ridge near Hendricks, Minnesota. Northern Alternative Energy's partners in the Minnesota project include Edison Capital, a subsidiary of Southern California Edison.

The world's largest single wind-power generation facility was dedicated September 26, 1998, near Lake Benton, Minn. Constructed, owned, and operated by Enron Wind Corp., the wind turbines of this facility are based on an earlier machine developed in cooperation with the U.S. Department of Energy. Electricity generated by this wind facility is sufficient to power 43,000 homes and will displace greenhouse gases equivalent to removing 50,000 new cars and light trucks from the road. Power is being purchased by the Northern States Power Company, with a service territory which includes much of Minnesota, as well as parts of Wisconsin, North and South Dakota, and Michigan.

Wind-power investment in the United States increased from $600 million in 1993 to $2 billion in 1996. Some third world countries are beginning use wind power to reduce a small measure of their reliance on fossil fuels. India, by the middle 1990s, had installed 900 megawatts of wind-powered electrical energy and had plans to develop 500 megawatts of solar (photovoltaic) energy. India also concluded an agreement for a joint venture with Enron and Amoco to build a 50 megawatt solar-energy plant that would provide electricity to about 200,000 homes.

Biomass Fuel

The manufacture of fuel from biomass has been proposed, mostly for ground transportation. In some corn-producing areas of the United States (one example is Nebraska), ethanol has been mixed with gasoline for several years. Such fuels have not done well in the competitive marketplace because their manufacture is from two to four times as expensive as oil-derived fuels. A fuel-cost comparison indicates that while gasoline could be refined for 15 to 16 cents per liter (in the late 1980s), the cost of biofuels ranged from an average of about 30 cents per liter (for methanol derived from biomass) to 63 cents per liter (for ethanol derived from beets in the United Kingdom).

As an attempt to reduce greenhouse gases, the growth and manufacture of biomass fuels has experienced problems other than cost of manufacture. Under some circumstances (if a biomass field replaced a forest, for example) this type of fuel might actually produce a net increase in emissions of greenhouse gases. Densely populated nations also would be hard pressed to spare from food production the large amounts of land required to raise vegetable matter destined for the world's gas tanks. When Brazil subsidized biomass production, for example, some farmers abandoned food because the government guarantees for biomass paid them more than they could earn for food on the open market.

Reforestation

Proposals have been made to ameliorate increases in greenhouse-gas emissions through reforestation, the purposeful planting of large forests to absorb some of humankind's surplus carbon dioxide. The 1997 Kyoto Protocol contains mechanisms whereby governments of countries such as the United States, which produce more greenhouse gases per capita than average, may earn credit toward meeting their emissions goals by subsidizing the preservation of forests in poorer nations.

Reforestation could help slow greenhouse warming, but only if trees are planted on a very large scale. For example, if an area of 465 million hectares was planted, the trees on this land area, once mature, would remove almost 3 billion tons of carbon dioxide from the atmosphere per year, or about 40 percent of the carbon that human beings add to the air. The creation of such a carbon sink would require a land area roughly half the size of the United States.

In England, some private firms have been planting trees to offset their contributions to global warming. The London Sunday Independent conducted a campaign during which readers bought more than 7,000 trees to offset the amount of carbon dioxide created by the manufacture of the newspaper over a year's time. The newspaper itself contributed 750 trees. The Glastonbury arts festival sold 1,333 trees to offset the equivalent amount of carbon dioxide produced by all the emissions created in the set-up, running, and dismantling of the show. The trade in trees is coordinated by an organization called Future Forests. Some musical groups, such as The Pet Shop Boys and Neneh Cherry, have produced 1.5 million "carbon-neutral" compact discs, meaning they have bought enough trees to offset the carbon emitted by production of their recording as well as movement of their stage materials around the country.

Ecologist George Woodwell estimates that one to two million square kilometers of newly planted trees would remove one billion tons of CO2 (1 ggt) annually, of the roughly seven million ggt being placed in circulation at the turn of the century. The problem would be finding large tracts of land fertile enough to support trees that isn't being used by human beings for other purposes. William R. Cline points out that reforestation is "a temporary remedy because a forest stores additional carbon only when it is expanding; once it reaches a steady state, the carbon released by dying trees often offsets that sequestered by new and growing trees".

A study of greenhouse-gas reduction potential in Chicago endorsed urban tree-planting projects as a way to reduce air pollution in cities, where reductions are needed the most.

A strategy focused on tree-planting would have a number of benefits which include CO2 absorption, removal of air pollution, and cooling and sheltering effects. Trees in Chicago have removed significant amounts of pollutants. In 1991, trees in Chicago removed an estimated 15 metric tons of carbon monoxide, 84 tons of sulfur dioxide, 89 tons of nitrogen dioxide, and 191 tons of ozone. Heating and cooling costs for buildings can also be reduced by strategically planting trees to shield buildings from wind and to provide shade in the summer months.

Photovoltaics and other Solar Energy Sources

Photovoltaic cells require no fuel (other than that provided by the sun) and little maintenance. The conversion of solar energy to electricity takes place with no moving parts and causes very little environmental disturbance. Because the sun must be shining to produce energy, however, it is likely that photovoltaic solar will develop, at least initially, with backup power from existing power-generation networks.

Photovoltaic cells already were cheaper than conventional, centralized power distribution to some remote locations, such as weather stations. This vast improvement in their efficiency probably will continue through the twenty-first century. By the end of the century, some present-day observers expect that many urban homes will produce their own electricity through photovoltaic solar panels mounted on rooftops. Centralized, fossil-fueled electricity generation may become obsolete, or one of several choices for electric-power generation.

During the 1970s, photovoltaic cells were manufactured by sawing large crystals of silicon into thin slices, an expensive and inefficient process. The cost of solar photovoltaic energy at the time was about $20 per peak watt. ("Peak," in solar energy terminology, means the cost of energy generated when the sun is at its peak elevation in the sky. The same word at a centralized, fossil-fueled or nuclear power plant means the cost of energy produced when customer demand is highest.) By the 1990s, however, work was underway to produce photo voltaics from semi conducter (computer) chips, potentially at a much lower cost once the technology is refined.

A study cited by Barry Commoner in his book Making Peace With the Planet indicates that aggressive expansion of photovoltaic technology could bring its cost down to $2-$3 per peak watt in one year, $1 per peak watt in three years, and 50 cents per watt in five years. At $1, photovoltaics would be cost-competitive with centralized electricity for roadside lighting. At 50 cents per peak watt, decentralized solar power could compete on a price basis with household power generated by fossil fuels in many U.S. markets. Gelbspan wrote in 1997 that by the middle 1990s, photovoltaic electricity was being produced at 3.2 cents per kilowatt hour (k.w.h.), and wind power at 3.0 cents per k.w.h., rates which could compete with coal or oil-fired electricity in some situations.

By the late 1980s, the Luz company of Southern California was selling enough solar-generated energy to Southern California Edison to meet the electricity demands of 270,000 homes. Technology also is being developed to produce "solar shingles," photovoltaic cells that can be positioned on the roofs of homes and businesses to generate electricity on-site. These are, perhaps, a harbinger of a new energy-delivery system, as well as a new source of power.

Solar power creates opportunities for decentralized "appropriate-scale" technology, especially in countries with large rural populations. One example is India, which averages 210 days a year of nearly direct sunlight, a large rural population, and a tradition of local, basic, small-scale problem-solving that stems from Mahatma

Gandhi, who turned homespun cloth from a small spinning machine into a powerful political symbol vis-à-vis a centralized weaving industry controlled by the British.

By 1995, six thousand villages in India that had no access to conventional power grids were drawing electricity from banks of photovoltaic solar cells. Using the same model of small-scale, locally controlled technology, photovoltaic modules and solar cooking stoves are being used increasingly in India's villages. Many villagers also use biogas digesters which convert the dung of cows and other animals to energy. The resulting methane is burned as energy before it bubbles into the atmosphere as a greenhouse gas. New technology also allows dung to be turned to an energy-rich sludge without smoke and fire. A million digesters were operating in India by 1990, despite the fact that one of them costs about $50 (with half the amount paid by a government subsidy), or almost one-fifth of the average rural Indian's annual cash income. The digester-financing program is administered by the Indian federal government's Department of Non-conventional Energy.

These programs should not leave an impression that India, as a whole, is reducing its greenhouse-gas emissions. In India's cities, a growing middle class is creating demands for more energy, most of it generated from fossil fuels, especially coal. India has only one percent of the world's coal reserves, but it is fourth among the world's nations in coal combustion. During the late 1980s, India had only 55,000 megawatts of electrical-generating capacity, twice the capacity of New York State, serving a population of more than 800 million. These figures suggest that electricity generation via fossil fuels is still in its infancy on the Indian subcontinent.

Injection of Carbon Dioxide into the Deep Oceans

The oceans are well known to scientists as a major "sink," or repository, for atmospheric carbon dioxide and methane. This fact has led to various proposals to remove the nettlesome oversupply of these gases from the atmosphere and inject at least some of it into the depths of the oceans.

Peter G. Brewer, Gernot Friederich, Edward T. Peltzer, and Franklin M. Orr, Jr. have demonstrated that deep-ocean disposal of carbon dioxide is technologically feasible. They described a series of experiments in Science during which carbon dioxide was lowered into several hundred meters of seawater, at different depths. The carbon dioxide, which is a gas at the surface, formed solid hydrates that were expected to remain in the ocean depths for "quite long residence times".

This idea has been dissected and dismissed by Hein J.W. de Baar of the Netherlands Institute for Ocean Sciences:

The crucial problem with fossil fuel CO2 is its very rapid introduction within 100 to 200 years into the atmosphere, as opposed to the very slow response of many thousands to millions of years of the deep ocean in absorbing such CO2. Eventually, the capacity for storage of CO2 in the deep ocean is very large. Yet, in the meantime, we will witness a transient peak of atmospheric CO2 which may yield catastrophic changes in the climate. Only after several thousands to millions of years most, but not all, of the fossil fuel CO2 will be taken up by the oceans.

In addition, according to Baar and Stoll, deep-sea carbon dioxide injection is suitable only for large stationary energy plants (30 percent of total human emissions) and would raise the cost of generation 30 to 45 percent, while decreasing efficiency by a similar percentage. Much of the carbon dioxide injected into the deep oceans would eventually return to the surface, doubling seawater's acidity, which would be toxic for fish, plankton, and other life in the oceans. "Deep-sea injection is at best a partial, expensive, and temporary remedy to the CO2 problem," write Baar and Stoll.

Atmospheric Modification

The mitigation strategies described in the 1991 NAS report conclude with consideration of atmospheric-modification projects. These include the placing of mirrored platforms in orbit to reflect sunlight, the use of guns or balloons to add dust to the stratosphere (to reflect sunlight), and the placing of billions of aluminized, hydrogen-filled balloons in the stratosphere, also to reflect incoming solar radiation. Additional strategies include the use of aircraft to maintain a dust cloud between earth and sun by making their engines less efficient in other words, intentional air pollution.

Another proposed solution along the same lines involves the burning of sulfur in ships and power plants to form sulfate aerosols, or particles. It is believed that the sulfur particles will stimulate the formation of sunlight-reflecting clouds over the oceans. Sulfur also has a well-known cooling effect in the atmosphere. Mikhail Budyko, a Russian climatologist, has proposed a massive atmospheric infusion of sulfur that would form enough sulfur dioxide to wrap the earth in radiation-deflecting thin, white clouds within months. The net effect, according to Budyko, would be to cool the earth in a fashion similar to the massive eruption of the volcano Tambora in 1815. The eruption of Tambora ejected enough sulfur into the air to produce, in 1816, an annual cycle known to climate historians as "the year without a summer." Crops across New England and upstate New York were devastated by frosts which continued into the summer. Mohawk Indians at Akwesasne, in far northern New York State, reported frosts into June.

Such atmospheric modification probably will remain an intellectual parlor game because the environmental costs of filling the stratosphere with sulfur dioxide (or other pollutants) far outweigh the benefits, even in a warmer and more humid world. The sulfur dioxide would have to be refreshed at least twice a month, as the previous load washes onto the earth, planet-wide, as acid rain. Sulfuric acid also tends to attract chlorine atoms, creating a chemical combination that could assist chloroflourocarbons (CFCs) in devouring stratospheric ozone.

Another instance of proposed intentional pollution involves fertilizing colder ocean waters with iron to stimulate the reproduction of oxygen-generating photoplankton. Tsung-Hung Peng believes that iron fertilization could lower carbon dioxide levels, but only by about 10 percent, under perfect conditions. J.H. Martin and R.M. Gordon also have been proponents of the iron-fertilization idea.

Yet another proposal involves the use of "lasers to break up CFCs in the atmosphere", an idea which evokes images of the Star Wars movies. Another proposal would create fields of photovoltaic solar cells on stations in orbit around the Earth or from bases on the moon. The tenuous nature of this technology is illustrated by a proposal that energy be transmitted from stations on the Moon to Earth via "some sort of energy beam". Little thought is given to the energy costs of transporting materials and human workers who would construct such stations, nor the energy cost of maintaining bases in space or on the Moon, much less the method by which the energy would be transmitted to consumers on Earth. At present, these proposals remain within the pie-in-the-sky, when-pigs-fly range of possibilities.

Eradication of Industrial Civilization

The German philosopher Friedrich Nietzsche once said, "The Earth is a beautiful place, but it has a pox called man". Bill McKibben's thesis, in The End of Nature, is similar: "We have built a greenhouse, a human creation, where once bloomed a sweet and wild garden". "If industrial civilization is ending nature, it is not utter silliness to talk about ending or, at least, transforming industrial civilization," writes McKibben. The "inertia of affluence, the push of poverty, [and] the soaring population" make McKibben pessimistic about humankind's chances of averting a general ruination of the Earth under greenhouse conditions.

John Gribben's Hothouse Earth, one of several popular treatments of global warming which appeared shortly after the notably hot summer of 1988, begins with a tribute to the philosophy of the Earth as Gaia, a living organism that will restore its ecological balance, even if restoring that balance requires extinguishing the human race. At the beginning of the book, Gribben quotes Jim Lovecock: "People sometimes have the attitude that 'Gaia will look after us.' But that's wrong. If the concept means anything at all, Gaia will look after herself. And the best way for her to do that might well be to get rid of us".

Gribben believes that what humankind is doing to the Earth's atmosphere will not be undone by human hands. His book ends with a profession that humankind will be unable to forestall the greenhouse effect by its own devices: "There is no prospect at all of bringing a halt to the release of carbon dioxide and other greenhouse gases, and thereby allowing the carbon cycles and the temperature of Gaia to return to normal".

Given rising populations, rising affluence, and rising levels of most greenhouse gases in the atmosphere, it is difficult, early in the twenty-first century, to conceive how the entire planet will be able to effectively achieve the technological and ideological paradigm shifts necessary to decouple human prosperity from the burning of fossil fuels. It is more difficult to imagine how our squabbling collection of nations and peoples will be able to respond as required to stop (much less reverse) an accelerating toward a warmer, more humid, and more miserable world. That said, a lot of authors have made retrospective fools of themselves by trying to forecast the future, so I can trust in my own fallibility, and hope, for the sake of the seventh generation, that my pessimism is at least partially in error.

12

Global Warming: Aerosols

Introduction

We discussed the possible climate effects of various factors, including greenhouse gases, variation in solar radiation and the earth's orbit, and the hydrological cycle. A powerful influence that has been largely ignored by climate modelers is that of atmospheric particles, or aerosols. The term aerosol refers, in the scientific sense, to very small particulate matter (liquid or solid), so small that it floats about on air currents for days, weeks, or even years at a time. Aerosols in this context are not to be confused with the spray-can aerosols in common household use. Household aerosols are usually solutions or powders sprayed by means of a pressurized gas. These sprays, composed of coarse particles that last for a short time before falling to the ground, are not a source for the atmospheric aerosols discussed here. The aerosols we discuss are from several sources, primarily volcanic eruptions, ocean dynamics, and human activities.

In early computer climate models, aerosols were given little or no attention because they were believed to have a minor or even negligible effect on global climate. The 1992 Intergovernmental Panel on Climate Change (IPCC) report reflects a major change in this attitude, giving greater attention to the cooling effects of aerosols because of both their direct reflection of solar radiation and their indirect participation in cloud formation. In fact, the importance of the aerosol cooling effect is recognized to be nearly equal in magnitude, but opposite, to the warming effect caused by the increase of greenhouse gases. This means that in the two years after the first IPCC report, the scientists on the panel found a global cooling process that equals the dreaded global warming.

The suddenness of this discovery of climatic effect due to aerosols is surprising because the phenomenon has been well known, if not well understood, since the early part of this century. In 1991 research groups from the Department of Atmospheric Sciences at the University of Washington and the Department of Meteorology at Stockholm University published a paper summarizing the early papers that pointed to a cooling effect caused by aerosols. In fact, one of these early papers was published in 1971 by Stephen Schneider, noted global warming scientist. Schneider's paper predicted global cooling based on anthropogenic aerosol effects.

Aerosol Composition

What are the qualities of these aerosols thought to cause global cooling? They are microscopic particles, or droplets, that exist for a relatively long time in the atmosphere because of their extremely small size. Formed from a multitude of sources, both natural and man-made, they have a varied and complex chemical makeup. Aerosols include silica dust particles from the desert, sodium chloride salt particles from the ocean, sulfuric acid droplets from burning of sulfur-bearing coal or oil, and sulfuric acid droplets originating from volcanic eruptions. Over the continents, a significant portion of particulate matter is formed by decaying biological matter, such as insects, bacteria, and pollen. And soot particles are formed from carbon wherever carbon-containing fuels such as coal, oil, wood, or paper are burned.

Each type of aerosol/particulate has its own interaction with incoming solar radiation and outgoing global radiation. These interactions depend on the aerosol's chemical composition, its particle number density, and the variation in size of its particles. The type and nature of the particles, as well as their location (height) in the atmosphere, are important to their atmospheric effects. Professor G. M. Krekov of the USSR Academy of Sciences defines five atmospheric altitude ranges used in developing mathematical models of aerosol behavior:

1. An earth surface boundary layer from the earth's surface to 2 kilometers (1.2 miles).
2. A turbulent mixing layer from 2-4 kilometers (1.2-2.5 miles).
3. The tropospheric background layer from the top of the turbulent mixing layer to the top of the troposphere, 10-12 kilometers (6-7.5 miles).
4. The stratosphere from about 12-30 kilometers (7.5 to ~19 miles).
5. The atmosphere above 30 kilometers (~19 miles).

Aerosols range in size from about 0.001 to 10 micrometers. (A micrometer is equal to one millionth of a meter.) To give you an idea of the sizes involved, visible cigarette smoke particles are about one to ten micrometers in size. The number density of particles at various locations in three different size ranges. Remember that this number density will vary with time, and these numbers are only averages. The actual number will fluctuate, depending on size distribution, winds, relative humidity, altitude, location, and a number of other factors. In general, the smaller the particles, the longer they float around in the atmosphere before they become part of a cloud or a raindrop or, in their random meandering, fall to the ground or into the ocean. The smallest particles are so small that they would not be visible under a conventional optical microscope. Particles up to about 0.1 micrometer last much longer than do larger ones; however, particles in the range of 0.1 to 1 micrometer are important as well. Particles larger than about 1 micrometer last only a short time in the atmosphere and are not considered important to long-term climate studies.

Table 12.1. Particle Number Concentration for Three Size Ranges and Various Locations

Location	Particle number concentration (number/cm^3)		
Size range and name	0.001-0.01ìm nucleation	0.01-0.1ìm accumulation	> 0.1ìm coarse
Polar	16.3	5.8	0.015
Background	10.3	41.9	0.077
Ocean	125	52.6	0.28

Remote continental	1260	81.1	0.44
Desert dust storm	1260	185	14.7
Rural	8610	1.7	0.34
Urban	135,000	1410	0.76
Stratosphere (20km)	0.4	4.1	0.018

Aerosol Effects

All aerosols reflect the sun's radiation like tiny mirrors, increasing the earth's albedo and providing a direct cooling effect to the earth's surface. The magnitude of this cooling effect will depend on the aerosol's number density, size distribution, altitude, and lifetime in the atmosphere.

There are additional interactions with radiant energy from both the sun and the earth. The nature of this interaction depends on the chemical composition of the aerosols and may lead to either a cooling or a warming effect. If the chemicals in the aerosol absorb the radiation that comes from either the sun or the earth, the effect can be atmospheric warming, but if the chemicals only reflect the radiation, the effect will be atmospheric cooling.

The very small particles can also act as cloud condensation nuclei, which are specks of microscopic dust on which water vapour can condense and contribute to cloud formation. Clouds generally have a cooling effect, but they can also cause warming under certain circumstances. Aerosols will initiate cloud formation when the humidity is right, the aerosols are in the right size range, and they reach the right altitude. This cloud formation will occur regardless of the source of the particles.

Remember that over 60% of the world is covered by clouds at any one time. In his 1994 Nature article, "Dirty Clouds and Global Cooling," Graeme Stephens, of the Department of Atmospheric Sciences at Colorado State University, summarized the growing evidence that clouds that are "dirty" from man-made aerosols provide a greater cooling effect than other clouds. The evidence indicates that there are more, but smaller, particles in dirty clouds, which are more efficient at reflecting the sun's radiation. Therefore, these particles increase the albedo of the polluted clouds. Stephens states: "When all effects of anthropogenically produced aerosol are taken together, then the predicted global forcing is comparable in magnitude to the direct forcing by greenhouse gases." This means simply that computer climate models can predict a cooling effect of greenhouse gases that is equal to the predicted warming effect! Stephens goes on to point out that the effects of aerosols are primarily regional and the global effects are subtle. All these cloud related factors complicate the overall aerosol effect on the global climate. The dynamics and subtle effects of aerosols on the fluctuating cloud cover of the globe have been inadequately dealt with in computer climate models.

Sulfuric Acid Aerosols

It is important to distinguish between the aerosols in the stratosphere and those in the troposphere. Aerosols in the stratosphere, the layer of atmosphere above the troposphere, arise from different sources. They must be treated differently by climate modelers from those in the troposphere because of their distinct chemical composition, circulation dynamics, radiational interactions, and many other factors.

The primary aerosol affecting the climate in the stratosphere is a sulfuric acid aerosol that consists of about 75% sulfuric acid (H_2SO_4) and 25% water, according to James Rosen of the University of Wyoming and Vladimir Ivanov of the Main Geophysical Observatory St. Petersburg, Russia, in their excellent chapter "Stratospheric Aerosols."[7] Nearly all sulfur-bearing aerosols become sulfuric acid if they reach the stratosphere. These aerosols come from both natural and anthropogenic sources, but most that reach the stratosphere come from volcanoes, according to D. J. Hofmann. Hofmann spent over twenty years studying atmospheric aerosols by means of balloon-borne instrumentation at the University of Wyoming before joining NOAA's Climate Monitoring and Diagnostics Laboratory at Boulder, Colorado.

Aerosols in the troposphere usually consist of a mix of the various particle types, such as sulfuric acid, soot particles, or insect parts. The tropospheric aerosols have a significant anthropogenic component, most of which comes from sulfur dioxide emissions associated with combustion of fossil fuels. Sulfur dioxide gas interacts with solar radiation and water vapour to form sulfuric acid aerosol.

Volcanoes

For many years the occurrence of a cold summer following a large volcanic eruption has been attributed to the emission of volcanic aerosols. Perhaps the first person to make this observation was Benjamin Franklin, following the great Laki eruption in Iceland and the Asama eruption in Japan in 1783. There are historical references to years with no summer and to years with summer snow in areas with typically very warm summer weather. But it has been only in modern times that scientific studies of volcanic emissions have provided the beginnings of a real understanding of the atmospheric effects of volcanic activity.

Table 12.2. Summary of Sulfur Containing Emission Sources

Source	Chemical form	Tg per year
Volcanic eruptions	mainly SO_2	7-10
Oceans	DMS	10-50
Soils and plants	DMS & H_2S	0.2-4
Biomass burning	SO_2	0.8-2.5
Anthropogenic emissions	SO_2	70-80

The chemical composition of the gases and particulates emitted by volcanic eruptions varies with the geology of the location. The major gas is sulfur dioxide (SO_2), but hydrogen chloride (HCl) and hydrogen fluoride (HF) gases are also emitted in large quantities, and there are many other chemicals emitted in smaller quantities. Rosen and Ivanov estimate that volcanoes produce from two to ten million tons per year of SO_2, about two million tons per year of HCl, and about 100,000 tons per year of HF. These gases are ejected explosively by the volcanic eruption, which means that volcanoes eject a great deal of sulfur dioxide gas directly into the stratosphere in addition to the dust and gas that remain in the troposphere. In the stratosphere the SO_2 is chemically changed into a sulfuric acid aerosol that can reflect sunlight back into space and thus directly cool the earth.

In the past, scientists thought that volcanic aerosols had short lifetimes, maybe a few weeks, and that therefore they had no major effect on the long-term global climate. However, during the past twenty or so years, measurements were made of the amount and size of aerosol particles in the stratosphere, using balloon-borne particle counters sent aloft from a facility at the University of Wyoming. These measurements indicate the buildup and decay of volcanic aerosols in the stratosphere. The increase of aerosol particles in the stratosphere when the volcanic event occurs is impressive, but more importantly, the slow decay of the numbers of these particles indicates a much longer effect than scientists had previously thought. The University of Wyoming is directly in the path of the Mt. St. Helens plume, which accounts for the large increase in particulates measured after its eruption.

Mt. Pinatubo was one of the three largest volcanic events of the twentieth century. It ejected about 3 cubic kilometers of material, which is about six times that of El Chichon, and it spewed forth about three times more sulfur dioxide (SO_2). The SO_2 cloud circled the earth in about three weeks and spread into both hemispheres. It was such an impressive event that Geophysical Research Letters, an important research journal for climate scientists, devoted a special section in 1992 to the early findings about the eruption.

Papers in this special section included a very interesting contribution from Jim Hansen and a group of researchers from NASA Goddard Space Flight Center. Hansen performed global temperature experiments using his computer climate model with an input of SO_2 of twice that emitted by El Chichon. The results of his model predicted a decrease in temperature (cooling) of about 0.5°C (0.9°F), occurring in the 1991-1993 period. Measurements of SO_2 from Mt. Pinatubo suggest that there could be three times the emissions as compared to El Chichon. If this is true, Hansen's computer climate models would have indicated additional cooling.

Hansen was careful to point out the many uncertainties and possible errors involved in the computer models in his paper, as most scientists do in their technical publications. For example, Hansen states that "a key mechanism which could limit the response to a negative climate forcing cooling, heat exchange with the deep ocean, is simulated very crudely in our model." He goes on to say that he needs more information about the nature of the sulfuric acid aerosol size distribution and the effects of the aerosol on the ozone layer. He is not sure if his model can handle the injection of water vapour into the stratosphere, and "climate feedbacks, such as changes of cloud properties and atmospheric water vapour, may not be accurately simulated in our climate model." Nonetheless, Hansen's prediction of 0.5°C cooling from the Mt. Pinatubo eruption is widely quoted in the scientific literature, including the 1992 IPCC report.

Hansen states that the effects of "Pinatubo aerosols should provide an acid test of climate models." This means that he feels the climatologists should be able to detect these effects in the climate and thus verify the computer climate models. Only time will tell.

There is a natural level of sulfuric acid aerosol in the stratosphere due to the ongoing volcanic activity on earth. According to Hofmann, there is evidence that this natural background in the stratosphere is increasing. If this is the case, climate scientists have yet another perturbation, or factor, that is unaccounted for in their models. The fact that the sun's radiation is decreased substantially after a large volcanic eruption is direct experimental evidence of the cooling effect.

Our knowledge of volcanic activity is incomplete; we can predict neither when nor how large a volcanic eruption will be. However, it is well known that the earth is geologically quite active, with its large, continental plates of earth mass continually pressing against each other, trying to relieve the planet's internal stresses. The actions of plate tectonics cause many smaller active volcanoes as well as the occasional major one. During the ten years between 1974 and 1984, for example, there were over seventy volcanic eruptions classified as "large." An earth that is producing over seven large volcanic eruptions per year may have a condition of naturally increasing stratospheric aerosol background, a condition that should be included in any computer global climate model.

As stated above, the aerosol component of the stratosphere is believed to be essentially from natural sources (mainly volcanic eruptions), with only a hint from anthropogenic sources. On the other hand, it would appear that much of the aerosol in the troposphere may be anthropogenic. Sulfur aerosols that do reach the stratosphere have a much larger direct climate effect than those contained in the troposphere.

Ocean-Derived Aerosols

Aerosols are produced at the ocean surface in a variety of ways. White caps caused by windy weather consist of millions of bubbles that burst into tiny droplets, many of which are so small that they are carried into the atmosphere. An aerosol of sea salt consisting mostly of sodium chloride (the same chemical as table salt) results. These salt particles are generally coarse and short-lived and do not affect long-term weather patterns. However, if global temperature does rise and the ocean winds increase, as predicted by the computer climate model of J. Latham and M. H. Smith of the University of Manchester in England, the increase in wind velocity will decrease the size of the sea salt aerosols. This decrease in size will create additional cloud-forming nuclei and increase the ocean clouds and the resultant cloud albedo (radiation reflection) to the point of offsetting global warming. These predictions have been derived from experimental measurements of the size distributions of sea-salt aerosol as related to the wind velocity. Whether this prediction of increased winds will prove accurate must wait on the "predicted" global warming; however, it presents another factor that computer models will need to include to be valid. Latham and Smith did not include the effects of the sea-salt aerosol on the dimethyl sulfide that occurs naturally from the oceans in their models.

Phytoplankton, the fundamental core of the ocean's food chain, generate a sulfur-bearing gas, dimethyl sulfide (DMS), as a product of their respiratory activity. The extent of DMS and its effects on cloud formation and the climate was the focus of a review article in Nature by a group of scientists, including Robert Charlson at the University of Washington and James Lovelock of Coombe Mill Experimental Station at Cornwall, England. They relate that DMS gas is produced by a number of different phytoplankton types. In addition, algae secrete sulfur compounds that react in the ocean to produce even more DMS. The DMS is short-lived in the atmosphere, with the final chemical product being sulfuric acid. Thus, sulfuric acid aerosols are indirectly generated in the oceans, and they account for most of the naturally occurring sulfuric acid aerosol in the troposphere.

Charlson 's paper concludes that there is little indication that the sulfuric acid from ocean aerosols ever reaches the stratosphere, but there is evidence that this source of sulfuric acid aerosol participates in cloud formation and contributes to increased cloud albedo in the warmer ocean regions near the equator.[21] This, of course, means that the ocean-generated aerosol may have a larger cooling role than has been thought and represents yet another factor not included in most computer climate models.

Anthropogenic Sulfur Emissions

Atmospheric sulfur emissions directly related to human activities make up well over half of all sulfur emissions. Most of these emissions are related to burning sulfur-bearing fuels or to processing sulfur-bearing ores or minerals to obtain base metals such as iron and copper. They are directly related to the industrialization of countries, and, according to Nicholas Lenssen, on the staff of the Worldwatch Institute, the developing countries are increasing these emissions at a much greater rate than the industrialized countries. These emissions, along with carbon dioxide and other greenhouse gases that are fundamental to the quality of life, are related to the size of the world's population.

It is not clear whether the global cooling effects due to the resulting sulfuric acid aerosols will offset the warming effects due to greenhouse gases. There will be believable arguments on both sides of the issue. However, it is clear that humans are responsible for emitting gases that cool the earth along with gases that warm it.

Regional Sources and Effects

There is a much larger concentration of particulate matter in the atmosphere over the continents than over the oceans or at the poles. This is especially true of very small particles, the size that can act as cloud condensation nuclei. There are from 10 to 1,000 times as many small particles per cubic centimeter over the landmasses as over the oceans. About 1,000 of these particles are due to dust in remote barren areas, as indicated by the "remote continental". There is an increase to nearly 9,000 particles per cubic centimeter in "rural" areas, presumably due to farming activities. The major source is clearly associated with the large, dense populations of the world's cities, however, which produce well over 130,000 particles per cubic centimeter.

In the remote and rural areas, particulate matter consists mostly of very small silica particles of common sand, along with decaying insects and other bits of biological matter. In rural agricultural areas the soil is exposed to winds between crops. The necessity of plowing the fields before planting crops increases the soil-containing aerosol over these regions. Furthermore, it is common practice in many parts of the world to burn off vegetative growth between plantings. Burning of any biomass creates carbon soot particles and sulfur dioxide gas, in addition to carbon dioxide and water vapour.

Enter the cities and you find automobiles, incinerators, electric power generation plants, factories, refineries, and other chemical processing plants. These sources emit various kinds of gases and particulates. The most prominent particulates are sulfate particles originating from SO_2 gas emitted by fossil fuel combustion. Some of these are large and last only a few hours or days in the atmosphere before arriving on someone's dresser or bookshelf as "ordinary dust." But many of the smaller particles will float with the winds and stay in the atmosphere for weeks. Regardless of size, during its residence time in the atmosphere, each particle acts as a mirror that directly reflects the sun's radiation. Two important considerations must be taken into account with respect to these aerosol sources. The first is that the major aerosol concentration is associated with the urban centers of the world. The second is that nearly all urban centers are in the Northern Hemisphere, so the primary cooling effect from these anthropogenic aerosols will be more prevalent in the Northern Hemisphere. More complication is thus added to computer global warming models.

Recent Studies

A 1992 Science article written by a group of climate scientists led by Robert Charlson from the University of Washington, with scientists from several other organizations, including Jim Hansen, estimated that tropospheric aerosol cooling equals the warming caused by CO_2 emissions. It is important to understand that this paper addresses cooling due to aerosols in the troposphere, whereas the paper that Hansen wrote about sulfur dioxide emissions from

Mt. Pinatubo was concerned with the cooling effects of the aerosol in the stratosphere. These are two different cooling effects.In recent years articles in the scientific literature have described at least four climatic effects from aerosols:

- Stratospheric cooling caused by volcanic injection of SO_2.
- Direct tropospheric cooling caused in part by man's burning of fossil fuels and industrial processing of metals.
- Enhanced cooling caused by "dirty" clouds.
- Potentially enhanced ocean aerosol generated by the increased winds associated with global warming.

Each of these cooling effects is purported to be equal in magnitude to the global warming caused by greenhouse gas emissions. If we add up all these cooling effects, we should be concerned but not about global warming. All these effects have not yet been included in computer climate models. It will be interesting to see the results of predictions when everything is included. Which will it be-global warming or global cooling?

Bibliography

Abrahamson, Dean Edwin. 1989. "Global Warming: The Issue, Impacts, Responses." In Dean Edwin Abrahmson, ed., *The Challenge of Global Warming*. Washington, D.C.: Island Press, 3–34.

Ackerman, A.S., O.B. Toon, D.E. Stevens, A.J. Heymsfield, V. Ramanathan, and E.J. Welton. 2000. "Reduction of Tropical Cloudiness by Soot." *Science* 288 (12 May): 1042–1047.

Ackerson, M.D., E.C. Clausen, and J.L. Gaddy. 1993. "The Use of Biofuels to Mitigate Global Warming." In Richard A. Geyer, ed., *A Global Warming Forum: Scientific,Economic, and Legal Overview*. Boca Raton, Fla.: CRC Press, 475–485.

Almendares, J., and M. Sierra. 1993. "Critical Conditions: A Profile of Honduras." *Lancet* 342 (4 December): 1400–1403.

Asakai, K. (2000) "Japan's opening statement to COP6," November 13, http://www.mofa.go.jp/policy/global/environment/warm/cop/cop6_9.html

Asuka, J. (1999) "Transfer of environmentally sound technologies from Japan to China," *Environmental Impact Assessment Review* 19: 553-567.

Bamber, Jonathan L., David G. Vaughan, and Ian Joughin. 2000. "Widespread Complex Flow in the Interior of the Antarctic Ice Sheet." *Science* 287 (18 February): 1248–1250.

Barnett, T.P. 1984. "The Estimation of 'Global' Sea Level Change: A Problem of Uniqueness." *Journal of Geophysical Research* 89:7980–7988.

Barnola, J.M., D. Raynaud, H. Oeschger, and A. Neftel. 1983. "Comparison of CO2 Measurements By Two Laboratories on Air from Bubbles in Polar Ice." *Nature* 303:410.

Barnola, J.M., D. Raynaud, Y.S. Korotkevich, and C. Lorius. 1987. "Vostok Ice Core Provides 160,000-year Atmospheric Record of CO2." *Nature* 329:408–414.

Carnell, R.E., C.A. Senior, and J.F.B. Mitchell. 1996. "An Assessment of Measures of Storminess: Simulated Changes in Northern Hemisphere Winter Due to Increasing CO2." *Climate Dynamics* 12:467–476.

Casten, Thomas R. 1998. *Turning Off the Heat: Why America Must Double EnergyEfficiency to Save Money and Reduce Global Warming*. Amherst, N.Y.: Prometheus Books.

Cruz, N.A. (2001) "CC impacts on water resources," in J.T. Villarin (ed.) *Disturbing Climate*, Quezon City: Manila Observatory.

Daley, Suzanne. 1999. "Battered by Fierce Weekend Storm, Western Europe Begins an Enormous Cleanup Job." *New York Times* (International Edition), 28 December, A-10.

Dalupan, M.C.G. (2001) "Policy and legal responses to the challenges of CC," in J.T. Villarin (ed.) *Disturbing Climate*, Quezon City: Manila Observatory.

Dansgaard, W., J.W.C. White, and S.J. Johnsen. 1989. "The Abrupt Termination of the Younger Dryas Climate Event." *Nature* 339:532–534.

Dansgaard, W., S.J. Johnson, H.B. Clausen, D. Dahl-Jensen, S. Gundenstrup, C.U. Hammer, C.S. Hvidberg, C.S. Steffensen, and J.P. Sveinbjrnsdottir. 1993. "Evidence for General Instability of Past Climate from a 250-kyr [thousand-year] Ice-core Record." *Nature* 364:218–220.

Depledge, J. (2000) *Tracing the Origins of the Kyoto Protocol: An Article-by-Article Textual History*, Technical paper prepared under contract to FCCC, August 1999/August 2000, FCCC/TP/2000/2, November 25.

ESSC (1999) *Decline of the Philippine Forest*, Quezon City: AEC Graphics.

Flavier, J.D.A., Baylon, M.L.L., and Paraso, G.R.V. (2001) "CC and public health in the Philippines," in J.T. Villarin (ed.) *Disturbing Climate*, Quezon City: Manila Observatory.

Forsyth, T. (1999) *International Investment Climate Change Energy Technology for Developing Countries*, The Royal Institute of International Studies. London: Earthscan.

Foundation for the Philippine Environment (FPE-1) (2001) *Hotter Facts on Hot Climate* (Primer on Climate Change 1), Quezon City.

Foundation for the Philippine Environment (FPE-2) (2001) *Cool Practices for Hot Climate* (Primer on Climate Change 2), Quezon City.

Gallares-Oppus, P.I. (2001) "Local climate action: fine-tuning local mindsets," in J.T. Villarin (ed.) *Disturbing Climate*, Quezon City: Manila Observatory.

Government of Japan (2000) *Japan's Comment on 'Financial Additionality' on CDM*, Mechanisms Pursuant to Articles 6, 12 and 17 of the Kyoto Protocol Principles, modalities, rules and guidelines for the mechanisms under Articles 6, 12 and 17 of the Kyoto Protocol Additional Submissions from Parties. Note by the secretariat. Addendum. FCCC/SB/2000/MISC.4/ Add.2. No. 8, Japan.

Lantin, R.S. (2001) "Philippine agriculture in a changing climate," in J.T. Villarin (ed.) *Disturbing Climate*, Quezon City: Manila Observatory.

Ministry of Foreign Affairs (MOFA) (1997) "Initiatives for sustainable development toward the 21st century, the Kyoto Initiative, assistance to developing countries for combating global warming summary,"

Muller, B.C. (2001) "Philippine policymaking and the FCCC," in J.T. Villarin (ed.) *Disturbing Climate*, Quezon City: Manila Observatory.

Narisma, G.T., Villa-Real, A.V., and Villarin, J.T. (2001) *Greenhouse Gases from Local Communities* (An Inventory Manual), Quezon City: Manila Observatory.

Nordic Council (1997) *Criteria and Perspective for Joint Implementation: Ten Nordic Projects in Eastern Europe*, Helsinki: TemaNord Energy.

Oberheitmann, A. (2000) "Possible conflicts of AIJ, JI and CDM Projects with provision governed by the WTO," *Joint Implementation Quarterly*, April.

Oberthur, S. and Ott. E.H. (1999) *The Kyoto Protocol: International Climate Policy for the 21st Century*, Berlin: Springer.

Orr, M. R. (1990) *The Emergence of Japan's Foreign Aid Power*, New York: Columbia University Press.

Overseas Economic Cooperation Fund (OECF) (1998) "Public opinion on environmental cooperation," (in Japanese), *OECF News letter* 60, Tokyo: OECF.

Perez, R. (2001) "Responding to the challenges of the rising sea," in J.T. Villarin (ed.) *Disturbing Climate*, Quezon City: Manila Observatory.

Tanabe, T. (1999) *Global Warming and Environmental Diplomacy: Battle at Kyoto Conference and development after* (in Japanese), Tokyo: JiJi tsushinsya.

Tibig, L.V. (2001) "Responding to the threats of CC to Philippine agriculture," in J.T. Villarin (ed.) *Disturbing Climate*, Quezon City: Manila Observatory.

UNCTAD (1998) Concept paper prepared by UNCTAD secretariat for the ad hoc international working group on the CDM organized by UNCTAD and UNEP.

Villarin, J.T., Narisma, G.T., Reyes, M.S., Macatangay, S.M. and Ang, M.T. (1999) *Tracking Greenhouse Gases: A Guide for Country Inventories*, Quezon City: Manila Observatory.

Yasutomo, D.T. (1986) *The Manner of Giving: Strategic Aid and Japanese Foreign Policy*, Washington, DC: Heath and Company.

Yonemoto, S. and Trindel, R. (1998) "Environmental diplomacy, regional security, and the limits of 'green aid'," *Asia-Pacific Review* Fall/Winter 5(2): 55-74.

Index